# Contest for Land in Madagascar

# African Social Studies Series



VOLUME 31

*The titles published in this series are listed at brill.com/afss*

# Contest for Land in Madagascar

## Environment, Ancestors and Development

*Edited by*

Sandra J.T.M. Evers
Gwyn Campbell
Michael Lambek

BRILL

LEIDEN · BOSTON
2013

*Cover illustration*: Villages near the COFAV (Corridor Fandriana Vondrozo) New Protected Area. Photo by Mijasoa Andriamarovololona, 2012.

This publication has been typeset in the multilingual "Brill" typeface. With over 5,100 characters covering Latin, IPA, Greek, and Cyrillic, this typeface is especially suitable for use in the humanities. For more information, please see www.brill.com/brill-typeface.

ISSN 1568-1203
ISBN 978-90-04-25622-4 (paperback)
ISBN 978-90-04-25623-1 (e-book)



This book is printed on acid-free paper.

CONTENTS

# LIST OF ILLUSTRATIONS

*Soodyall, Morar and Jenkins*

*Figures*



*Map*



*Raharijaona and Kus*

*Figures*



*Table*



*Campbell*

*Figures*

*Maps*



*Tables*



*Anderson*

*Figures*

### *Middleton*

*Map*



### *Walsh*

*Map*



### *Seagle*

*Table*

# LIST OF CONTRIBUTORS

THOMAS ANDERSON is a Lecturer of History at the University of New Hampshire. He specializes in World and Environmental Histories. His current research explores the environmental history of Madagascar as well as the role of naturalists in the Indian Ocean.

GWYN CAMPBELL, Canada Research Chair in Indian Ocean World History at McGill University, was born in Madagascar and is the author and editor of many works, including *An Economic History of Imperial Madagascar* (Cambridge, 2005) and *David Griffiths and the Missionary "History of Madagascar"* (Brill, 2012).

SANDRA J.T.M. EVERS is associate professor in Anthropology at VU University Amsterdam. Her research and publications deal with migration, slavery, cultural transmission; natural resource management, poverty and sustainable development in Madagascar. Dr Evers directs a research programme on foreign large-scale land acquisitions in Africa and Madagascar.

Emeritus Professor TREFOR JENKINS (MD, London) held the position of Head of Department of Human Genetics from 1975 until his retirement in 1998. He is currently an Honorary Professorial Research Fellow in the Division of Human Genetics.

MICHAEL LAMBEK is Professor of Anthropology and Canada Research Chair at the University of Toronto, Scarborough. His books include *Knowledge and Practice in Mayotte* (Toronto 1993) and *The Weight of the Past: Living with History in Mahajanga, Madagacar* (Palgrave Macmillan 2002).

SUSAN KUS is Professor of Anthropology at Rhodes College in Memphis. Her research and publications deal with early state organization in central Madagascar with particular interest in state ideology in societies of primary orality and the symbolic organization of space and landscape.

KAREN MIDDLETON specialises in the social anthropology and environmental history of southern Madagascar. Her published work on kinship, ritual, and ecology combines long term fieldwork in the region with extensive archival research.

BHARTI MORAR (PhD, Wits) is the Laboratory Manager for the Centre for Clinical Research in Neuropsychiatry in Perth, Western Australia. She has conducted population genetics research on the Malagasy and the European Roma (Gypsies).

VICTOR RAHARJAONA is an independent researcher affiliated with the Université de Fianarantsoa. His research and publications focus on early state organization in central Madagascar, orality and oral traditions, and ethnographic work with Betsileo ritual specialists (*mpanandro*) who orient houses and tombs.

CAROLINE SEAGLE (MSc, VU University Amsterdam) is a PhD candidate in anthropology. She is co-editor (with Sandra Evers and Froukje Krijtenburg) of *Africa for Sale? Positioning the State, Land and Society in Foreign Large-Scale Land Aquisitions in Africa* (Brill, 2013).

HIMLA SOODYALL (PhD, Wits) is a Medical Scientist in the Division of Human Genetics at the National Health Laboratory Service and Associate Professor at the University of the Witwatersrand, South Africa.

VENUSIA VINCIGUERRA has researched conflicts arising from exploitation of land and natural resources in West Africa and Madagascar. She has a MA in Peace and Conflict Studies from the University of Oslo, and a BA from the University of L'Aquila.

ANDREW WALSH is an Associate Professor of Anthropology at the University of Western Ontario. His research concerning Madagascar's sapphire and ecotourist trades is documented in *Made in Madagascar: Sapphires, Ecotourism, and the Global Bazaar* (University of Toronto Press, 2013).

# LAND COMPETITION AND HUMAN-ENVIRONMENT RELATIONS IN MADAGASCAR

Sandra J.T.M. Evers, Gwyn Campbell, and Michael Lambek[1]

*Ny tany vadiben-Janahary: mihary ny velona, manotrona ny maty*
("The earth is God's chief wife: she maintains the living and guards the dead")
Malagasy proverb (Houlder 1929, pt.1: 6).

On 17 March 2009, after a bitter three-month power struggle, a military-backed coup toppled the incumbent Marc Ravalomanana and replaced him with a hitherto unknown 34-year-old ex-disc jockey named Andry Rajoelina as President of Madagascar. Rajoelina, during his prior two year tenure as mayor of Antananarivo, claimed that Ravalomanana had put up Madagascar "for sale" on the international market, pointing to a purported land deal with Daewoo Logistics which would have earmarked approximately 1,3 million hectares of land to palm oil and maize (Teyssier et al. 2009/2010, Vinciguerra in this volume). He sensed that the land issue was a tinder box he could ignite through a populist campaign which could serve as justification for a seizure of power. He further charged that Ravalomanana's one-dimensional economic development agenda failed to offer a social plan likely to improve the lot of the population (Burnod 2009; Randrianja 2012), over 70 per cent of whom live below the poverty threshold (World Bank 2012; Sarrasin 2006: 389).

Rajoelina's meteoric rise to power and politicisation of the land issue marked a fundamental turning point in Madagascar's history. Despite being a home-brewed revolution, the Daewoo affair dragged the country into the ideological arena of 'land-grab' politics, which already involved

---

[1] This work results from research and a conference generously funded by the Canada Research Chairs Fund, the Social Sciences and Humanities Research Council of Canada, the Beatty Memorial Lectures Committee (McGill University), the Maxwell Cummings Distinguished Lectureship Committee, the Provost's Office (McGill University), the Dean, Faculty of Medicine (McGill University), Department of Human Genetics (McGill University), the International Institute for Advanced Studies (McGill University), and the Indian Ocean World Centre (McGill University). We also thank the authors and peer-reviewers of this volume for their valuable feedback. We are grateful to Natalja Laurey for her assistance in formatting this manuscript.

investors ranging from Wall Street, China and India to international conservationist groups. Whether Rajoelina's intentions were local or wider in scope, his campaign ensured that Madagascar would henceforth be front and centre of international debates on land and environmental issues that highlighted the island as variously a biodiversity hot-spot, zone of neo-colonialism, and 'green grab' victim. In the swirl of this hot political debate, there was a real risk that polemic would gain the ascendancy over discussion.

These events underscore the significance of this volume, which aims to examine land as a subjective, ontological reality for Malagasy people within the context of the historical, economic and cultural factors influencing the way they view land. For the Malagasy, large land acquisitions, endemic poverty, health and sovereignty loom large, as they do for many other peoples, notably on the African continent. However, land for the Malagasy is of particular importance as it also straddles the boundaries between the here and now, and the hereafter. This is what makes Madagascar significant as a terrain to understand the extent to which international and local treatments of land can be problematic. The assessment of multinational mining companies, state corporations acting as proxies for their respective governments, hedge fund companies or conservation groups is very much rooted in realpolitik and the achievement of policy aims over clear timelines. It is the aim of this book then to examine a more comprehensive range of human-environment interactions within the increasingly fierce climate of competition for land.

This book tracks key aspects of human-environment interactions, particularly focusing on the contemporary dynamics of World Bank-inspired land titling and other land projects. By 2009, over three million hectares of Madagascar's arable land was subject to foreign direct investment (FDI) aimed at promoting tourist, mining, infrastructural and agricultural projects (Uellenberg 2009). The 1999 enactment of new mining legislation has opened Madagascar to large-scale mineral investment (Sarrasin 2006). Despite Rajoelina's rhetoric and the abandonment of Daewoo's agricultural venture, the number of high-impact land projects in the country, particularly in mining (Andrianirina – Ratsialonana et al., 2011; Evers et al. 2011) have grown. Such developments were accelerated by the de-regulation of the Malagasy economy, guided by World Bank policies to open up the Malagasy economy for foreign investment while simultaneously offering Malagasy rural dwellers incentives to invest in land through offering them 'tenure security' (see Evers and Middleton chapters). NGOs started up and funded land titling projects throughout

Madagascar. In this volume, we discuss mining investments (including their partnerships with conservation NGOs) and land titling for local Malagasy, which have been at the core of land issue debates and policies in Madagascar over the past few decades.

Nature conservation and climate mitigation schemes can also be viewed as large-scale land acquisitions (Cotula et al. 2009; Smaller and Mann 2009; IIED 2009). These acquisitions are often characterized as 'green grabbing', or the appropriation of land and resources for environmental purposes (Fairhead et al. 2012). Conservation zones in the island have increased in size following Ravalomanana's 2003 pledge to triple the size of protected areas to six million hectares. Simultaneously, climate mitigation in the form of forest protection (referred to as REDD: Reducing Emissions from Deforestation and Degradation of Forests) has been on the rise in the island (Ferguson 2009). Forest management is becoming increasingly subject to external pressures due to the active engagement of conservation NGOs imposing restrictive access policies in forest areas (c.f. Duffy 2006, Harper 2002, Horning 2008, Walsh 2005, Ferguson 2010). However, many Malagasy depend on forest products like wood and food items (GISC 2009), pitching them against these NGOs and foreign mining companies interested in exploiting the subsoil minerals of forestland.

We also re-examine whether and to what extent human-environment interactions have impacted on Malagasy ontology as a result of these developments. Swyngedouw (2004) and Whatmore (2002) observe that human relations are embedded in the environment or landscape, where land underlies and conditions the entire belief system. Samuels (1979), has observed that landscapes are formed in constant dialogue with social, cultural and economic factors and expressions, Roymans et al. (2009) developed the concept of 'landscape biography' in order to consider how landscapes are shaped as an outcome of a long-term and "complex interplay between the history of mentalities and values, institutional and governmental changes, social and economic developments and ecological dynamics (p. 339)." Landscapes are always transformed over time, re-shaped, re-used and re-moulded according to changing conditions. As landscapes or features of landscapes change, the social space – meaning imbued in a landscape – changes as well.

World Bank policies, foreign investments and conservation projects have wrought major changes in Malagasy relations to their land and social and natural environment. Conventional dichotomies pitting 'local' against 'foreign', or 'traditional' versus 'modern' mores have proved insufficient as tools to analyse how the Malagasy position themselves when competing for land. In order to properly identify the key issues of human-environment interactions in the island's history, we propose commencing by a review of the meaning of land and forests for the Malagasy against the backdrop of World Bank policies concerning land titling and the stimulation of land investment.

### *The Merging of the Local and Foreign in Malagasy History*

The origin of the first settlers of Madagascar has been a subject of considerable speculation, some of it scholarly, and some framed in terms of racial and Orientalist bias. European commentators from the sixteenth century noted the presence of two main groups in Madagascar, whom they distinguished as 'Negroid' and lighter-skinned—described variously as Malay or even Chinese. The 1614 Peter Kaerius map refers to 'white' natives of Madagascar: "supposed to have been transplanted out of China" (Wheatley 1975: 76). Scholars have also proposed that Arabs, Babylonians, Phoenicians, Greeks, Romans and Jews may have visited and possibly settled in the island in ancient or historical times (Grandidier 1902: 3; Vérin 1986: 26, 28). Indigenous traditions are generally vague: while the 'Islamic' dynasties on the southeast coast whose origins go back to the fourteenth century, claimed an origin in Mecca (Vérin 1986: 80–9), and some Bara groups point to an African heritage (Campbell, unpublished fieldwork 1993), other Malagasy simply provide the explanation that their forefathers came from "overseas". Given the prevalence of bilateral descent, such accounts of foreign ancestry could also refer to a handful of in-marrying individuals rather than of whole groups (Lambek 2002).

The dual 'African' (e.g. zebu cattle) and 'Southeast Asian' (e.g. rice terraces) material culture affords some credence to the theory of mixed African and Southeast Asian origins for the Malagasy population. However, it is generally accepted that by the 1880s all Malagasy, regardless of their physiognomy and culture, spoke variants of an Austronesian language (from the Indonesia-Melanesia-Australasia-Polynesia region). This singular fact formed the basis for the theory expounded by the father and son team of Alfred and Guillaume Grandidier that all Malagasy originated from the 'Malayo-Polynesian' region, the ancestors of 'Negroid' peoples being dark-skinned 'Negritos' from Melanesia. The Grandidiers argued that from about the mid-first millennium BCE successive groups of these adventurous mariners sailed their outrigger canoes directly across the Indian Ocean to Madagascar – a distance of about 4,800 km – as contemporaneously their kin were colonising the Pacific (Grandidier 1908;

see also Kobishchanow 1965: 137). Unfortunately, such theories did not yet recognize the point subsequently established by Franz Boas that biological traits, language, and material culture are not intrinsically connected to each other and can travel or diffuse independently of one another (Boas 1966).

The Grandidiers' viewpoint became widely accepted so that the conventional French and consequently Malagasy intellectual view was that the Malagasy were of Malayo-Polynesian (Austronesian) 'stock'. However, as the chapter in this volume by Himla Soodyall, Bharti Morar, and Trefor Jenkins shows, recent advances in DNA analysis, and their application to the Malagasy, clearly demonstrate that the first settlers of the island were of both Austronesian and African origin (see the second chapter of this volume).

We have intentionally opened our discussion with an inquiry into the origins of the Malagasy, to recall the fact that the Malagasy roots originate outside the island which considerably muddies distinctions between 'local' and 'foreign'. Among others, Anna Tsing cautions against over-simplifications of 'local' and 'foreign' – which (in the same vein as Mosse 2005 and Appandurai, 1996) she glosses as 'global'. She even notes this tendency in her own work: "I find myself doing it. Yet we know that these dichotomies are unhelpful" (2005: 58). She calls this process 'scale-making' and warns that concepts like 'local' and 'global' should not be taken at face-value:

> I argue that scale is not just a neutral frame for viewing the world; scale must be brought into being: proposed, practiced, and evaded, as well as, taken for granted. Scales are claimed and contested in cultural and political projects (ibid.).

Homogenization processes of the 'local' and 'global' risk masking the diversity of practical strategies deployed by the Malagasy in the defence and negotiation of their land claims. The characterisation of such land projects as simply capitalist schemes crushing Malagasy land users overlooks the active roles played by the Malagasy in the construction of socio-economic configurations that shape their interactions with migrants and other perceived outsiders.

### *The History of the Environment and Land Competition*

When the ancient continent of Gondwanaland broke up, Madagascar became detached from the east coast of Africa and remained in isolation from human contact until around the BCE/CE changeover. This isolation resulted in the development of a significant number of unique species (Madagascar has been termed a 'floating museum' because of its rich biodiversity). The island currently possesses what may be the world's greatest concentration of unique flora and fauna: 98 percent of Madagascar's land mammals, 92 percent of its reptiles, 68 percent of its plants, and 41 percent of its indigenous bird species are found nowhere else. Discoveries are still being made of new species on the island—615 between 1999 and 2010 alone (Mustain 2011).

Publications on Madagascar's environmental history often fail to mention that the protection of Malagasy forests dates further back than the colonial period. In 1794, King Andrianampoinimerina implemented the first national decree against "slash and burn" forestry (Wright, 1997: 384). In 1881, Queen Ranavalona II imposed the "code of the 305 articles" prohibiting tree burning or cutting in order to prioritise forest conversation. Such policies were implemented to develop an economy based on permanent rice fields rather than shifting horticulture and livestock production – which was seen as harmful to the environment (Goedefroit 2007: 17). Human-environment relations were clearly of concern in precolonial times, although during the French colonial period a more systemic environmental legislation appeared.

The academic consensus is that in the pre-1800 era, climate probably had a greater impact than humans on the Malagasy environment but that with increased European intervention in the island, notably during the commercial boom that accompanied the development of the international economy in the nineteenth century, and under the French colonial regime from 1895, human activity had a dramatically increased impact (Esoavelomandroso 1979; Vérin 1990; Brown 2000; Campbell 2005). In his chapter, Tom Anderson describes how, during the nineteenth century, Madagascar became 'orientalised' in the eyes of Europeans, as they made the discovery of its unique natural habitats. He also notes how Madagascar, once feared as a strange and dangerous enclave, evolved into an environment that could be mastered by science. Anderson charts the process by which science became a tool for control of land and resources, and culminated in the introduction of the first nature reserves (*Réserves Naturelles Intégrales, RNI's*) into the country during the 1920s.

European reports of the island's unique flora and fauna did little to slow down the exploitation of its forests and other natural resources (*infra*). Thus, under Radama I (r. 1810–28) and Ranavalona I (r.1828–61), Europeans such as Marius Arnaud /Arnoux (1809–29), Julien Gaultier de Rontaunay

(1783–1863), and Napoléon de Lastelle (1802–56) established plantations and negotiated export monopolies on the east coast; while the American commercial agent William Marks acquired similar monopoly prerogatives for exports on the west coast (Campbell 2005: 99–101, 166, 170–1).

However, these foreigners were heavily regulated by the imperial Merina regime, the policies of which, as Gwyn Campbell shows in his chapter, resulted in the destruction of far more woodland and primary forest than that caused by either foreigners or peasant farmers. The imperial Merina era, 1790–1895, was characterised by the adoption of policies designed to promote autarkic economic modernization, notably in the production of armaments, in order to deter European political intervention. Campbell examines how, in the absence of domestic capital formation, the Merina court exploited Madagascar's human and natural resources: Merina subjects, as well as peoples conquered by Merina armies, were conscripted into *fanompoana*, a system of unremunerated and compulsory forced labour for the state, and the island's natural resources were exploited in attempts to promote both a domestic industrial revolution and exports of raw, semi-processed and manufactured goods. However, the human and environmental consequences of such policies contributed towards undermining the Merina regime, and have indelibly marked the subsequent history of Madagascar (ibid. and Campbell's chapter in this volume).

Possibly the main human impact upon the environment prior to the eighteenth century was the extinction of the island's largest animals (pygmy hippopotamus, giant tortoise, giant lemurs, and Aepyornis – a giant flightless bird); and destruction by fire of savannah belts, bush and woodlands in the centre and east to create agricultural plots and in the south and west to create pasture. Although the causes of these transformations remain undetermined, there is increasing evidence that climactic rather than anthropogenic factors may have been the primary factors leading to reduced woodland (Dewar 1997; 2007). Over time, the primary human impact on the environment has unquestionably been due to rice cultivation. Rice growing predominated on the fertile mud flats and riverside marshes on the east, southeast and west coasts, where seeds were scattered. It also dominated agriculture in the central highlands. There, seeds were initially scattered in marshland, but from the early eighteenth century were increasingly planted, and nursery shoots replanted. Rice was also the primary crop along the eastern escarpment, where swidden ('slash-and-burn') was practised. Secondary crops included sweet potatoes, manioc (cassava), maize, haricot, sugar cane, *saonjo* and *tavo* (edible arums), banana, plantain, sorghum, and *voanemba* (a species of bean) (Campbell 2005: 23–9). The development of terraced and irrigated rice fields in Imerina and Betsileo necessitated enormous inputs of labour, but with two harvests annually resulted in heightened output, and an increase in population growth and concentration by the late eighteenth century that led to extensive forest clearance on the plateau for firewood and, increasingly, for pasture (Dubois 1938: 429; see also Campbell's chapter).

During the French colonial era, policy from Paris was predicated upon the idea that the colony should pay its own way and provide tropical products for the metropolis. As of 1896, the French colonial administration offered liberal land grants to French metropolitan and Mascarene Creole settlers and companies to establish cash crop plantations, and exploit the island's cattle, mineral, and forest resources (de Coppet 1947; Cros 1923; Deschamps 1972: 248–52, 256–8, 281–96). In 1904, the French company Suberbie, which from 1886 had obtained from the Imperial Merina court a gold mining concession at Maevatanana, also moved into graphite mining – which boomed during the First World War (Campbell 2005: 103–5; Deschamps 1972: 292). The colonial administration invested in a transport and communications network: 14,500 km of road by 1933; 876 km of rail by 1936 designed to enhance the extraction and exportation of resources (Deschamps 1972: 283). Both private and public concerns were responsible for the substantial elimination of the eastern tropical rainforest where the main transport lines transected, connecting the highland interior to the main east coast ports; and where the plantation frontier notably for coffee, the main export cash crop, continued to expand (de Coppet 1947. vol.2: 124–44; Campbell 2006: 67–99). Thus, paradoxically, policies and practices leading to both the destruction and conservation of Malagasy forests were evident in the colonial period. Hence contemporary developments in mining and in conservation partnership projects (see Seagle in this volume) were well anchored in the history of the island.

Population growth is another factor to be considered when contemplating human-environment relations. While Madagascar experienced largely stagnant demographic growth for most of the nineteenth and early twentieth century when the population fluctuated between two and three million people, it started to expand rapidly from the mid-1920s. The population increased from 3.4 million in 1921 to 3.7 million a decade later, to reach 4.1 million in 1941 (Campbell 2005: 137; de Coppet 1947. vol.2: 75), which in turn would become a factor in increased competition over land.

### *Contemporary Perceptions of the Environment and Competition for Land*

Susan Kus and Victor Raharijaona's contribution to this volume contains an essential discussion of the meaning of *ny tany*. *Ny tany* in its primary sense is 'land, soil, the earth', but can also be used to designate '*pays, contrée, royaume*' (1955, 673 dictionary of Abinal and Malzac), terms that carry connotations of empire and state involvement in land issues. The 'land, the earth' is a powerful metaphor for leadership in Merina kingdoms in which "harvest and famine become indexes of governance". As witnessed by the events of 2009, land continues to play a powerful role in contemporary political dynamics in the island. Rajoelina's portrait of Ravalomanana as having put Madagascar's land up "for sale" implied both treason and a condemnation of Ravalomanana's poor governance (cf. Vinciguerra's chapter).

For many Malagasy, the human (*olona*) includes both the living and the dead (the ancestors), the relations between whom is anchored by land. Land inevitably mediates cultural meanings, knowledge complexes, symbols and ontologies, but also involves labour systems, human health and human relations (cf. Evers & Seagle 2012). Such perspectives provide insight into how, theoretically, one might approach Malagasy perceptions of the environment and the 'ethos' of life itself (Keller 2008) – wherein dichotomies between nature and culture are somewhat artificial. The Malagasy possess a profound religious, socio-political and economic attachment to land. Firstly, land anchors not only the individual, but also entire kinship groups to their history, for it connects the living with the dead via the family tombs on the ancestral land (see Bloch 1971; Feeley-Harnik 1991; Middleton 1999). It thus lies at the heart of Malagasy cultural life. Malagasy people revere the ancestors, whom they believe to be possessed of powers that they use to exert influence over the lives of the living. Consequently, the Malagasy take considerable pains to remain on good terms with the ancestors. This desire is expressed in a multitude of practices concerning the building and renewal of tombs and the care of corpses, including, in some regions, festive practices of reburial or rewrapping of corpses, and in others of maintaining relics from important ancestors (see e.g. Bloch 1971; Feeley-Harnik 1991; Middleton 1999; Ballarin 2000; Baré 1980; Lambek 2002). The tomb is the portal to the hereafter, the world of the ancestors, serving as the passage through which the deceased undergo 'ancestralization'. In this ritual process, the deceased are transformed into ancestors and become a source of renewed life. Moreover, it is the ancestors who legitimize rulers, so that throughout Malagasy history, monarchs and leaders had to reckon with ancestral traditions (see Raharijaona and Kus chapter). Simultaneously, one should be cautious not to exoticise Malagasy relations with land and the wider environment as, in the quest for survival, some Malagasy engage in activities that deviate from such perspectives. Reports of poaching and trafficking in unique species and rosewood are frequently reported in national and international news media. It is likely that one local group's sacred ancestral land is viewed as subject to exploitation by another group or members of a different class.

Such reports fuel conservationist concerns that one of the world's "biodiversity hotspots" – an expression term first coined by Conservation International ecologist Norman Myers – is under threat. Myers' wish was to identify places rich in biodiversity but increasingly threatened by environmental degradation and, ultimately, species extinction. In *Biodiversity Hotspots for Conservation Priorities* (2000: 853), Myers et al. ten tropical forest regions were identified as biodiversity hotspots – a term now applied to 34 regions worldwide. In total, those hotspots account for 16 percent of the planet's surface area and 50 percent of the world's endemic species (Neimark and Schroeder 2009). Madagascar, one of the top priority zones, is discussed by Thomas Anderson in this volume who notes that, because of its unique ecosystem, Madagascar has been central to some of the most important scientific debates, from Darwinism to biogeography and human origins (see also Feeley-Harnik 2001).

The declaration of Madagascar as environmental hotspot certainly opens the gates to increased environmental intervention in Madagascar. Building from Richard Grove's work (1995), Tsing (2000) provides an explanation for the construction of global environmental science (see also Anderson's chapter) and how the latter shaped what is today known as a "universal environmental knowledge":

> ...the imperial placement of scientists in botanical gardens and research stations across the European colonies inspired continent-crossing correspondence in the late 18th century. Through this correspondence, informed by widespread fears of climate change caused by colonial deforestation, colonial scientists formulated notions of a 'global' climate. This commitment to planet-wide environmental process allowed further developments in imagining both science and policy on a global scale (Tsing 2000: 348).

Conservation NGOs commonly express widespread fears of climate change and call for urgent measures in response. Many articles, images, graphs and statistics about land clearance on the "Red Island" indeed look alarming: 90 percent of Madagascar's forest cover is gone, 200,000 ha of its primary forest are disappearing each year, and its endemic species, which represent 80 percent of all mammals, reptiles, plants and amphibians on the island, are living on "borrowed time" (Harper et al., 2007: 325; Hannah et al. 2008). Marion Desmurger (2012) observes the fact that while Madagascar is on the one hand presented as "unique and threatened", on the other conservationists posit it as a place that needs to be saved for the sake of universal conservation. This has the direct consequence that "... playing with space in such a way that Madagascar is both 'localised' and 'globalised' can lead to a situation whereby the environment becomes detached from the cultural and the spatial, that is, from people and land" (p. 103). So, Malagasy issues on the ground become 'global' concerns and through 'scale-making' of Malagasy as destructors of the 'global commons' conservationist intervention is legitimised.

The growing impact of human activity has widened the debate on how best to protect Madagascar's unique natural environment; the Malagasy government has come under increasing international pressure to restrict human infringement on remaining areas of natural forest and other habitats rich in indigenous flora and fauna, notably from agencies such as the World Wildlife Fund (WWF), Lemur Conservation Foundation, Primate Conservation Inc., Madagascar Fauna Group, Wildlife Trust, Conservation International, Missouri Botanical Garden, Wildlife Conservation Society, US Peace Corps, and World Bank, which in 2000, for example, inaugurated a programme to increase the planting of pine and eucalyptus to satisfy fuel needs. Such programmes increasingly go under the banner of REDD – Reducing Emissions from Deforestation and Forest Degradation (Angelsen 2009). Kanowski et al. (2011) stress that REDD+ (including benefit sharing) has the potential to help reduce poverty and improve rural livelihoods. REDD+ is said to enhance previous Payment for Ecosystem Services (PES) schemes, limited to small geographical areas, by its nationwide scope (Tomimura 2012). Madagascar was among the first 14 countries to benefit from the World Bank's REDD+ policies (World-Bank 2008), with an estimated 1.75 million ha of forest projects being targeted for REDD projects (Rakotoarijaona 2010). These projects were initiated before the *coup d'état* but continued after the take-over by Rajoelina. Conservationists promoting such land schemes commonly start from the premise that poverty is a major reason for infringement on forest land and other areas rich in natural life. However, whether poverty alleviation has been achieved by such conservation programmes in Madagascar is far from certain (See e.g. Hanson 2007, Harper 2002, Walsh 2005).

Madagascar is also the setting for the acquisition by international companies of large tracts of land, like elsewhere in Africa, for industrial-scale cultivation. This phenomenon is frequently labelled as 'land grabbing', or as a neo-colonial manifestation of capitalist forces intent on subjecting Africa to their economic and political interests. South African farmers have been acquiring land in Zambia, Mozambique and Tanzania for decades, but on a relatively small scale. By contrast, recent acquisitions have been large-scale, in the form of 50 to 99 year leases negotiated by externally based organizations and governments. Some of these acquisitions are by Western investors: for example, in 2008, the UK company CAMS acquired 45,000 ha in Tanzania to cultivate sorghum for biofuels; and in 2009 the US company Jarch Capital acquired 400,000 ha of farmland in South Sudan. Other external investors come from lands across the Indian Ocean, notably India, China and the Middle East. In December 2008, Varun Agriculture, an offshoot of Varun Industries, a global conglomerate based in India, acquired 231,912 ha in Madagascar for food production. This new "scramble for Africa" has been sparked chiefly by growing food security problems in investor countries, concerning insufficient domestic food supply due to limited availability of water and arable land; higher production costs; bottlenecks in storage and distribution; expansion of biofuel production (a competing land and crop use); increasing urbanisation; and changing diets (Cotula et al 2009).

Most countries affected are poor and located in Sub-Saharan Africa. These foreign land investments therefore are often promoted as a "development opportunity". The question however remains as to what kind of development they would facilitate and who stands to benefit from them. While governments tend to welcome this foreign direct investment (FDI) in land, such contracts have been widely censured on a number of grounds. Firstly, the leases create considerable local tension as they conflict with customary tenure (only 2–10 percent of land in Africa is held under formal land tenure) and usage, and are for periods of up to 99 years. Secondly, they frequently result in the forcible eviction and displacement of small-scale cultivators, often without promised compensation or permanent paid employment by the foreign investor on the same land (Rice 2010; Vidal 2010). Certainly, the Malagasy government has found it difficult to effectively regulate large international concerns. Among other factors, long-term leases allow them considerable flexibility for land exploitation without a high level of accountability. In the example of the Varun's project, of the four crops to be grown – rice, wheat, maize and pulses – only in the case of rice (60 percent) did the Malagasy government stipulate that more than 50 percent of output go to the domestic market; while 100 percent of pulses were designated for export (Anon 2009).

The emphyteotic (50–99 years) leases granted to foreign concerns mean that, in a country with an average life expectancy of 63.6 years (Index Mundi 2011), few dispossessed cultivators can aspire to reclaim their land within their lifetime. The spectre of renewal amounts to exclusion of successors and heirs to ancestral lands. Demographic pressures on land is increasing at a rapid rate: whereas average population density in Madagascar was 7 per $km^2$ in 1950, it reached 26 per $km^2$ in 2000, and is estimated to rise to 73 per $km^2$ by 2050 (Cotula et al 2009: 60). Some commentators consider that FDI could liberate the African cultivator from the crushing poverty caused by customary tenure and traditional land use (See e.g. Sanders 2010). However, that prognostic is of little solace to small Malagasy farmers for whom land is the major source of livelihood but also much more than an economic consideration.

Many environmentalists and politicians view security of tenure as an important incentive to small-scale farmers to invest in a long-term and sustainable use of land. For example, conservationists who identify 'slash-and-burn' agriculture in Madagascar, within the context of rapid demographic growth, as a major contributory factor to biodiversity loss, have long argued for formal codification of land rights. Those who reject the premise of Malthusian population-induced degradation in favour of the 'Boserupian' hypothesis of agricultural intensification driven by increasing land scarcity, also consider secure land title to be a contributing factor for such intensification (See e.g. Goldman 1993).[2]

In 2004, the Malagasy government under Ravalomanana launched the *Programme National Foncier* (PNF) creating a national land registration system which was to serve as a platform for further land investments and forest conservation measures. In 2005, the Malagasy government signed a Foundation for Protected Areas and Biodiversity, supported by co-signatories Conservation International and the World Wildlife Fund; and in December 2005, the first extra 10,000 $km^2$ of the new Protected Areas System of Madagascar were granted protection status. However, as Sandra Evers discusses in her chapter, registration has proven to be problematic due to the competing notions posited by customary law, anchored in the ancestral concepts of land, and positive law, inspired by the French Civil Code. Land registration has placed Malagasy customary land access arrangements and *vazaha* (foreign) economic and cultural values on a collision course, further exacerbated by existing and emerging social

[2] Karen Middleton, written communication 14 April 2009.

hierarchies within Madagascar. The Malagasy draw upon both customary law (*lex loci*) and positive law (*lex fori*) to secure their land access. Sandra Evers and Karen Middleton argue in their chapters that the focus on such opposing concepts should not ignore the hierarchical structure of Malagasy social configurations and unequal land access arrangements which are antipathetical to the notion that land registration could provide tenure security.

In her chapter, Middleton adds that a dualistic understanding of land tenure ('modern' versus 'traditional') obscures the true complexity of issues in the region she studies – Karembola (dryland Madagascar). Noting that the notion that peasant farmers automatically invest in their land once they secure title – a cornerstone of World Bank-endorsed titling schemes – does not always hold in the Malagasy context (cf. Jacoby & Minten 2007), she argues for a more nuanced understanding of land tenure in Madagascar. From her longitudinal ethnographic research carried out between 1991 and 2003 she demonstrates that individuals often have overlapping but unequal access rights to land, and suggests that, rather than phrase the problem in terms of competing jurisdictions, it is relevant to examine local constructs of 'traditional' or 'customary' law and 'modern' state law as co-evolutionary over time. In one community, interactions between local practices of cactus-planting to claim ownership and state legislation on tree planting and land tenure, combined with perceptions of climate change and growing pressure on land, have led many small farmers to clear more indigenous forest than was needed for subsistence production in order to plant exotic prickly pears. In the second case study, tenure rules and practices, while equally contested within local communities, have been shaped by other factors, including exclusions relating to Cap Sainte Marie Special Reserve and community disputes around an invasive prickly pear. Highlighting the complex factors to be considered when modelling correlations between agricultural intensification, biodiversity conservation, and land titling, Middleton argues that land titling and the 'one-size-fits-all' approach may not be the panacea for Madagascar's problems that some assume.

Land pressure at the local level and competing claims from outsiders is indeed a leitmotif that appears throughout our discussion of contemporary land issues in Madagascar. Andrew Walsh in his study of a northern Madagascar region demonstrates how tenure relations are also affected by changed restrictions in access to land due to the establishment of a local national park. He shows how exclusively focussing on park policies risks ignoring village dynamics of land access. In a fascinating ethnographic account, he relates how social relations are affected when migrants come to the villages in the context of a sapphire rush (small-scale artisanal mining). Initial migrants (*vahiny*, 'visitors') follow customary practices of gaining land access, privileges granted in exchange for reciprocity relations of deference to small town leaders (*lehibe*), seen as 'fathers' of the place, and adhering to cultural requirements. However, as immigration increases, such rules erode and relations between village leaders and migrants degenerate. This is particularly the case of young immigrant miners who ignore the *lehibe*'s status as founding fathers and, when raising children, follow cultural practices that deviate from ideals of the *lehibe*. Through his study, Walsh emphasises that 'local' Malagasy are not a homogeneous group, demonstrating the heterogeneous nature of the *lehibe*, and the variety in their reflections and practices over time, which underlines the importance of examining the historical component of land issues in Madagascar.

Caroline Seagle in her contribution on mining-conservation partnerships in South East Madagascar also highlights the point that Malagasy assessments and practices of land are indeed closely tied in with social relations. Analysing the way in which two large-scale mining companies, Rio Tinto/QMM and Sherritt/Ambatovy, engage with biodiversity conservation and "offsetting" strategies, she builds the argument that "green" mining in Madagascar may not adequately account for complex, "illegible" land-based livelihoods, and may in turn uphold the nature/culture dialectic. Drawing upon theoretical approaches to embodiment and the materiality of human-environment interactions, she goes on to show how mining-conservation engagements create a "representational space" (Lefebvre 1991) wherein only some aspects of the social-environmental reality are mediated, while others are not. In doing so, she analyzes both the BBOP (Business and Biodiversity Offset Programme), of which Sherritt/Ambatovy is part, and Rio Tinto's "biodiversity book" – a synthesis of research carried out on the rare and biodiverse littoral forest set aside for strip mining, but which omits the company's own impacts while drawing upon degradation narratives to focus on local misuse of the environment. As "green" mining must reconcile the need for land both for mineral extraction and environmental protection (through "biodiversity offsetting" mechanisms), the risk remains for local Malagasy people to lose access to crucial natural resources.

These ethnographic accounts of competition for land help us to establish the background for the events of late 2008 and early 2009. In one of the world's largest land deals, the Malagasy government purportedly was on the verge of concluding an agreement to rent 1.3 million ha of land – roughly half of Madagascar's current arable land – on a 99-year lease to Daewoo Logistics Corporation of South Korea to cultivate maize and palm oil for export. This forms the theme of Venusia Vinciguerra's contribution in this volume (see the final chapter). Most external commentators have viewed popular Malagasy reactions to this deal as a protest against a neo-colonial system that would make them a "South Korean colony" (Ryall and Pflanz 2009). However, this was not the only factor feeding discontent. Many Malagasy, in addition to concerns they have due to increasing land competition, still have deep-rooted attachments to land as 'ancestral' that, as Middleton and Evers show, are highly rational even if (or precisely because) they are incompatible with capitalist systems and modalities.

It is within this context that one must view the current highly charged political and economic climate in Madagascar in which the dynamics of land are skewed by a multitude of stakeholders competing for access to and acquisition of land. For the Malagasy, self-identity, historical continuity, life, and value are anchored in the ancestral soil. To place a uniquely monetary value on land, and offer it for sale or lease, notably to foreigners, is seen as the ultimate sign of disrespect towards the living and the dead (the ancestors). As Lambek was frequently informed in Mahajanga, "the ancestors are not for sale." Given growing impoverishment, and a lack of investment in sectors of the economy that might afford them alternative employment opportunities, access to land is for many Malagasy their sole prospect of survival. Any consideration of the relations of Malagasy people to the land and their wider environment needs to take into account the conjunction of ecological, historical, political, economic, cultural, and religious contexts within which such relations have been formed and developed. Through exploring the past and present relations of Malagasy people to land, the essays in this volume lay the basis for understanding the complex issues that lie at the core of the current political and economic crisis in Madagascar.

## *Bibliography*

Abinal, A. and Victorin M. 1955[1888]. *Dictionnaire Malgache-Français*. Paris: Editions Maritimes & Coloniales.

Angelsen, A. ed., 2009. *Realising REDD+: National strategy and policy options*. Bogor, Indonesia: Centre for International Forestry Research (CIFOR).

Anon, 2009. Varun-Madagascar contract (23/03/2009) – http://farmlandgrab.org/2849 (accessed 24/07/11).

Appadurai, A. 1996. The production of locality. In: Richard Fardon, ed., *Counterworks: Managing the Diversity of Knowledge*. London: Routledge, pp. 204–225.

Andrianirina Ratsialonana, R., Ramarojohn, L., Burnod, P. and Teyssier, A. 2011. *After Daewoo? Current Status and Perspective of Large-Land Scale Acquisitions in Madagascar*, Rome: Observatoire du Foncier à Madagascar, CIRAD, ILC.

Ballarin, M. 2000. *Les Reliques Royales à Madagascar: Source de légitimation et enjeu de pouvoir (XVIIIe-XXe)*. Paris: Karthala.

Baré, J. 1980. *Sable Rouge: Une monarchie du nord-ouest malgache dans l'histoire*. Paris, L'Harmattan.

Bloch, M. 1971. *Placing the Dead*. London: Seminar Press.

Boas, F. 1966. *Race, Language, and Culture*. Glencoe IL. Free Press.

Brown, M. 2000. *A History of Madagascar*. Princeton, NJ: Markus Wiener.

Burnod, P. 2011. Appropriations foncieres: après l'affaire Daewoo, que se passé-t-il a Madagascar? CERI, CNRS (Juin 2011). http://www.ceri-sciences po.org/archive/2011/juin/dossier/art_pb.pdf, (accessed 26 November 2012).

Campbell, G. 2006. Coffee production in Madagascar. In Clarence-Smith, William Gervase and Topik, Steven. eds., *Coffee under Colonialism and Post-Colonialism: The Global Coffee Economy in Africa, Asia, and Latin America, 1500–1989*. Cambridge: Cambridge University Press: 67–99.

Campbell, G. 2005. *An Economic History of Imperial Madagascar, 1750–1895. The Rise and Fall of an Island Empire*. Cambridge: Cambridge University Press.

Campbell, G. 2004. Madagascar. In Krech III, Shepard, McNeill, J.R., and Merchant, Carolyn (eds.). *Encyclopaedia of Environmental History*. New York: Routledge. Vol 2: 796–798.

Cotula, L.,Vermeulen, S.; Leonard, Rebecca and Keeley, James, 2009. *Land grab or development opportunity? Agricultural investment and international land deals in Africa*. Rome: FAO.

de Coppet, M. ed. 1947. *Madagascar et Réunion* 2 vols. Paris: Encyclopédie de l'empire français.

Cros, L. 1923. *Madagascar pour tous. Comment aller. Que faire à Madagascar. 8.000 Français y sont déjà*. Paris : Albin Michel.

Deschamps, H.J. 1972. *Histoire de Madagascar*. Paris: Berger-Levrault.

Desmurger, M., 2012. Translating the sustainability transition: The creation and implementation of a translocal environmental project, from Paris to Madagascar. Research master thesis African Studies Centre, Leiden, Netherlands.

Dewar, R.E. 1997. Were People Responsible for the Extinction of Madagascar's Subfossils, and How Will We Ever Know? In Goodman, Steven M. and Patterson, Bruce D. eds. *Natural Change and Human Impact in Madagascar*. Washington: Smithsonian Institution Press: 364–77.

Dewar, R.E. and Richard, A.F. 2007. Evolution in the hypervariable environment of Madagascar. *Proceedings of the National Academy of Sciences* 104. 34: 13723–13727.

Dubois, H.M., 1938. *Monographie des Betsileo (Madagascar)*. Paris: Institut d'ethnologie.

Duffy, R., 2006. Non-governmental organisations and governance states: The impact of transnational environmental management networks in Madagascar. In *Environmental Politics*, 15:5. pp. 731–749

Esoavelomandroso, M. 1979. *La province maritime orientale du « Royaume de Madagascar » à la fin du XIXe Siècle (1882–1895)*. Antananarivo: FTM.

Evers, S.J.T.M, Burnod, R. Andrianirina Ratsialonana, A. Teyssier. 2011, Foreign land acquisitions in Madagascar: competing jurisdictions of access claims. In *African engagements: Africa negotiating an emerging multipolar world*. T. Dietz, K. Havnevik, M. Kaag, and T. Oestigaard, eds. Pp. 110–132. Leiden: Brill Academic Publishers.

Evers, S.J.T.M. and Seagle, C. (2012) Stealing the Sacred: Why "Global Heritage" Discourse is Perceived as a Frontal Attack on Local Heritage-making in Madagascar. *Madagascar Conservation & Development* (S2): 97–107.

Fairhead, J., M. Leach and I. Scoones, 2012. Green Grabbing: a new appropriation of nature? *Journal of Peasant Studies,* 39(2): 237–261.
Feeley-Harnik, G. 1991. *A Green Estate: Restoring Independence in Madagascar.* Washington, DC: Smithsonian.
Ferguson, B. 2010 *Bibliography of critiques of Malagasy conservation.* University of East Anglia, School of International Development.
Ferguson, B. 2009. REDD comes into fashion in Madagascar. *Madagascar Conservation and Development* 4(2): 132–137.
GISC 2009. Etude sur la Production et la Consommation en Produits Forestiers Ligneux à Madagascar, Version actualisée 2009, Report to IRG and USAID Madagascar.
Goedefroit, S. 2007. Quel Développement À Madagascar? *Etudes Rurales N° 178* Editions De L'école Pratiques De Hautes Etudes En Sciences Sociales – 24/07/2007.
Goldman, A. 1993. Agricultural Innovation in Three Areas of Kenya: Neo-Boserupian Theories and Regional Characterization. *Economic Geography* 69. 1: 44–71.
Grandidier A. 1902. *Histoire de la Découverte de l'Ile de Madagascar par les Portugais pendant le XVIe siècle.* Paris: C. Lamy.
Grandidier, A. and G. Grandidier. 1908. *Histoire Physique, Naturelle et Politique de Madagascar. Ethnographie de Madagascar.* vol.4. tomes 1–3. Paris: Imprimerie Nationale; tome 4. Paris: Hachette et Société d'Éditions Géographiques, Maritimes et Coloniales).
Grove, R. *Green Imperialism.* Cambridge: Cambridge University Press.
Hannah L, et al. 2008, Climate change adaptation for conservation in Madagascar. *Biol. Lett. 4, 590–594.*
Hanson, P. 2007. Governmentality, Language Ideology and the Production of Needs in Malagasy Conservation and Development. *Cultural Anthropology* 22(2): 244–284.
Harper, J. 2002. *Endangered Species: Health, Illness, and Death among Madagascar's People of the Forest.* Durham NC: Carolina Academic Press.
Harper, J. et al., 2007. Fifty years of Environmental Conservation deforestation and forest fragmentation in Madagascar. *Environmental Conservation* 34 (4): 1–9.
Horning, N.R. 2006. Strong Support for Weak Performance: State and Donors' Mutual Dependence in Madagascar, IASCP Draft (Prepared for the International Study of Common Property's 11th Bi-Annual Meetings Ubud, Indonesia June 19–23, 2006).
Houlder, J.A. 1929. *Ohabolana : or Malagasy Proverbs, Illustrating the Wit and Wisdom of the Hova of Madagascar* part 1. Antananarivo : Friends Foreign Missionary Association.
IIED, 2009. 'Land grabs' in Africa: can the deals work for development? (Briefing, September 2009).
Index Mundi, 2011. Madagascar Life expectancy at birth – http://www.indexmundi.com/madagascar/life_expectancy_at_birth.html (accessed 23/07/11).
Jacoby, H. and M. Bart, 2007. Is Land Titling in Sub-Saharan Africa Cost-Effective? Evidence from Madagascar. *The World Bank Economic Review* 21(3), 461–485.
Kanowski, P.J., C.L. McDermott, and B.W. Cashore, 2011. Implementing REDD+: lessons from analysis of forest governance. *Environmental Science and Policy* 14(2): 111–117.
Keller, E. 2008. The banana plant and the moon: Conservation and the Malagasy ethos of life in Masoala, Madagascar. *American Ethnologist* 35(4): 650–664.
Kobishchanow, Y.M. 1965. On the Problem of Sea Voyages of Ancient Africans in the Indian Ocean. *Journal of African History* 6. 2: 137–41.
Lambek, M. 2002. *The Weight of the Past: Living with History in Mahajanga, Madagascar.* New York: Palgrave MacMillan.
Lefebvre, H. 1991 [1974]. *The Production of Space* (transl. by Donald Nicholson-Smith). Oxford: Oxford University Press.
Middleton, K. ed. 1999. *Ancestors, Power, and History in Madagascar.* Leiden: Brill.
Myers, N., R.A. Mittermeier, C.G. Mittermeier, G.A.B. da Fonseca and J. Kent, Biodiversity hotspots for conservation priorities, *Nature* (403) 24 February 2000.
Mosse, D. 2005. *Cultivating Development: An Ethnography of Aid Policy and Practice.* London: Pluto Press.

Mustain, A. 2011. Madagascar offers treasure trove of new species (06/06/2011) – http://www.msnbc.msn.com/id/43296171/ns/technology_and_science-science/t/madagascar-offers-treasure-trove-new-species/ (accessed 23/07/11).
Neimark, B. and R. Schroeder 2009 Hotspot discourse in Africa: Making space for bioprospecting in Madagascar, *African Geographical Review* 28: 43–70.
Rakotoarijaona, J.R. 2010. Madagascar REDD readiness preparation progress. *In PCPF PC7*. Washington, DC.
Randrianja, S. 2012. *Madagascar, le coup d'État de mars 2009*. Paris; Karthala.
Rice, T. 2010. *Meals per Gallon: the impact on people and global hunger*. London: Action Aid.
Roymans, N. et al. 2009. Landscape Biography as Research Strategy: The Case of the South Netherlands Project. *Landscape Research* 34(3): 337–359.
Ryall, J. and Pflanz, M. 2009. Land-rental-deal-collapses-after-backlash-against-colonialism *Daily Telegraph* – http://www.telegraph.co.uk/news/worldnews/ africaandindianocean/madagascar/4240955/Land-rental-deal-collapses-after-backlash-against-colonialism.html (accessed 29/08/2011).
Sarrasin, B. 2006. The mining industry and the regulatory framework in Madagascar: Some developmental and environmental issues. *Journal of Cleaner Production* 14: 388–396.
Smaller, C, and H. Mann. 2009. A thirst for distant Lands: foreign investment in agricultural land and water. International Institute for Sustainable Development (IISD) (May 2009).
Swyngedouw, E. (2004), *Glocalisations*. Temple University Press, Philadelphia, PA.
Teyssier, A., L. Ramarojohn, and R. Andrianirina Ratsialonana. 2009/2010. Des terres pour l'agro-industrie internationale? Un dilemme pour la politique foncière malgache. *Echogéo*, 11.
Tomimura, C. 2012 The Future of Tropical Forests in the Tide of Global Climate Change. *Journal of Sustainable Forestry* 31(1–2): 174–183.
Tsing, A.L. 2000. The Global Situation. *Cultural Anthropology* 15(3). 327–360.
Tsing, A.L. 2005. *Friction: An Ethnography of Global Connection.*. US: Princeton University Press.
Uellenberg, A. 2009. *Foreign Direct Investment (FDI) in Land in Madagascar*. Eschborn: Deutsche Gesellschaft für Technische Zusammenarbeit (Work of Division 45 – Agriculture, fisheries and food, GTZ, Eschborn).
Vérin, P. 1990. *Madagascar*. Paris: Karthala.
Vérin, P. 1986. *The History of Civilisation in North Madagascar*. Rotterdam: A.A.Balkema.
Vidal, J. 2010. 21st-Century African Land Grab. *The Observer*.
Walsh, A. 2005. The Obvious Aspects of Ecological Underprivilege in Ankarana, Northern Madagascar. *American Anthropologist* 107(4): 654–665.
Whatmore, S. 2002. *Hybrid Geographies: Natures, Cultures, Spaces*. London: Routledge.
Wheatley, P. 1975. Analecta Sino-Africana Recensa. In Chittick, H.N. and Rotberg, R.I. Eds. *East Africa and the Orient*. New York: Africana Publishing.
World Bank. 2012. Madagascar – Country Brief. http://web.worldbank.org/WBSITE/EXTERNAL/COUNTRIES/AFRICAEXT/MADAGASCAREXTN/0,,menuPK:356362~pagePK:141132~piPK:141107~theSitePK:356352,00.html (Accessed 26 November 2012).
World Bank. 2008. First Countries Named to Benefit from Forest Carbon Partnership Facility; 14 developing and 9 industrialized countries formally join partnership to reduce emissions from deforestation and forest degradation (REDD). Press Release No. 2009/029/SDN.
Wright, P.C. 1997. The future of biodiversity in Madagascar: A view from Ranomafana National Park. In Goodman S.G. & Patterson B.D. (ed.). *Natural Change and Human Impact in Madagascar*. Washington D.C: Smithsonian Institution Press. pp. 381–405.

# THE GENETIC TRAIL TO MADAGASCAR

Himla Soodyall, Bharti Morar & Trefor Jenkins[1]

The focus of this volume is on human claims to land in Madagascar, and the shifting relationship over time between the island's peoples and the environment. These are complex issues and we are reminded of the "The Land Ethic" idea proposed by Aldo Leopold in which he argues that "There is as yet no ethic dealing with man's relation to land and to the animals and plants which grow on it" (Leopold 1987: 237). Leopold challenges us to view conservation as a state of harmony between humans and land with a broader understanding of ecological conservation (Leopold 1987: 243). According to Leopold, an understanding of ecology should embrace concepts from geography, history, botany, agronomy and economics (Leopold 1987: 261). We would contend, that an understanding of the origins of the people occupying the land is equally important. In this context, this chapter reviews how humans first came to occupy Madagascar. In other words, who are the Malagasy and what is their relationship to one another?

The prehistoric settlement of Madagascar is one of the great mysteries of human migration (Burney 2005). The significance of understanding prehistoric times has been recognised in recent years as an important element in planning for the future in terms of ecological and cultural restoration (Burney 2005)—both of which are largely shaped by human-environment interactions. Early French botanists painted a picture of the landscape of Madagascar as being densely covered by forests, and suggested that when the proto-Malagasy arrived, they proceeded to burn the landscape thereby transforming it to grassland and arid bushland

[1] We wish to thank all the individuals who volunteered to participate in our research. This research was approved by the Human Research Ethics Committee (MEDICAL) at the University of the Witwatersrand. We are also most grateful to Drs G. Campbell, T. de Ravel, P. Willem, A. Rakouth, and S. Quanbeck for their assistance with the collection of blood samples in Madagascar. This research was supported from grants awarded to T. Jenkins and H. Soodyall from the South African Medical Research Council, South African Institute for Medical Research (now the National Health Laboratory Service) and the University of the Witwatersrand, and the National Geographic Society (grant 5113–93) awarded to T. Jenkins. We are also most grateful to Mr David de Veredicis for his assistance with research associated with this manuscript.

(Humbert 1927). On the contrary, recent paleobotanical data suggest that the prehuman Madagascar was a mosaic of forest and environments including wooded savanna, open grassland, and semi-arid bushland in which prehuman fires and megafaunal herbivory were major factors (Burney et al. 2004).

Multiple independent lines of paleoecological research have suggested that people were present on the island around 2,300 years ago, but whether they settled then is still inconclusive from these data. For example, David Burney's review of botanical clues to human presence in Madagascar indicates that the first occurrence of pollen of hemp (*Cannabis* sp.) in sediments throughout the island was identified around this time (Burney 2005: 387). In addition, the record is consistent with decreasing amounts of pollen of woody species and an increase in ruderal (disturbance-adapted) pollen types, a large increase in microscopic charcoal particles in sediments, decline of dung-fungus spores, and earliest evidence of butchery of large animals that later become extinct.

Ethnobotanical data have also provided clues in tracing the geographic origin of the earliest Malagasy people. Early Asian plants introduced to Madagascar include hemp, which was thought to have been brought from Asia to Africa by Indonesian traders over two millennia ago, certain coconut varieties (*Cocos nucifrera* L.), the banana (*Musa acuminata Xbalbisiana* Colla.), and rice (*Oryza sativa* L.) (Burney 2005: 387). Fossil pollen also provides evidence of a subsequent human migration from Africa to Madagascar. The castor bean (*Ricinus communis* L.), an African plant valued for medicinal and insecticidal qualities, appears in the pollen record of Madagascar's highlands about 1000 years ago (Burney 1987: 137). Burney suggests that the re-appearance of *Sporormiella* spores, a likely indicator for livestock, in the same sediment probably signalled the advent of large-scale pastoralism in Madagascar. He further suggests that "...Islamized Indian Ocean traders, sometimes referred to as Swahili or Afro-Arab peoples, began colonizing the west coast of Madagascar about this time. These people would likely have brought castor beans and cattle-raising to the island" (Burney 2005: 387).

Human survival is dependent on the environment. It is therefore important, when exploring human-environment links in Madagascar in the context of conservation and development discourses, for scholars to examine (1) the nature and composition of prehuman environments; (2) when people arrived on the island, and who they were; (3) human subsistence and dependency on plants and animals; and (4) anthropometric landscape transformations and their consequences. In this paper, we explore further the second issue: The origin of the first human migrants to Madagascar, and the date of their arrival.

Despite the increasing wealth of linguistic, archaeological, paleobotantical, and historical data in reconstructing the past history of Madagascar, there remain many unanswered questions related to the peopling of the island. These include: Who were the first occupants? How did they get there? What is the origin of other parental groups who settled on the island? To what extent have these groups admixed? How have males and females contributed to shaping the present day gene pool? This paper makes use of genetic data to shed more light on some of these questions.

## *Genetic Studies and Historical Reconstruction*

Human population genetic studies have had a tremendous impact on assessing how the gene pool of living people has been shaped by past demographic and historical events. Firstly, the genetic approach in historical reconstruction is based on an examination of contemporary lineages found in living people and then using the known phylogeographic distribution of these lineages to infer their most likely origin. Secondly, since changes or mutations in a lineage occur in a stepwise manner, it is possible to reconstruct the history or evolutionary relationship of the divergent lineages back to a hypothetical common ancestor. Thirdly, using a calibration of a molecular clock for particular DNA segments, it is possible to date when certain clades (or groups of lineages) evolved, and how they relate to the most common recent ancestor (MCRA).

From the early 1990s, an inter-disciplinary team of researchers from the University of the Witwatersrand in Johannesburg, including historian Gwyn Campbell and medical professionals and human geneticists Trefor Jenkins, Thomy de Ravel, Pascal Willem and Himla Soodyall, initiated a collaboration with Jean Roux, Director of the Pasteur Institute in Antananarivo at the time, and pathologists Stan Quanbeck and Andriamanantsoa Rakouth of the Malagasy Lutheran Church Health Department (SALFA). Between May 1992 and November 1996, approximately 2300 blood samples were collected from individuals throughout the island, with their informed consent. Our preliminary research made use of samples collected from individuals in whom both their maternal and paternal ancestry was derived from the same ethnic group. We discuss herein how haploid DNA markers [mitochondrial DNA (mtDNA) and Y chromosome DNA] have contributed in tracing the genetic trail(s) to the geographic region of origin of the mtDNA and Y chromosome DNA found in living people in Madagascar. In addition, we examine the use of a few autosomal DNA markers in assessing population affinity and gene admixture in the Malagasy.

## *Genetic Studies on the Malagasy*

Before 1990, few genetic studies focussed on elucidating the origins of the Malagasy. Ronald Singer and colleagues, using blood groups and the haemoglobin S (sickle cell trait), revealed that one-third of the genetic heritage of the islanders could be traced to Indonesian origins and the remaining two-thirds to African origins (Singer et al. 1957). They also found that the sickle cell trait occurred in the Malagasy at a frequency of 4.9%. John and Vina Buettner-Janusch, employing a better method to examine the sickle cell trait (starch gel electrophoresis versus the use of reducing agents used in the former study), refined the estimate for haemoglobin S to 10.7% (Buettner-Janusch and Buettner-Janusch 1964); while René Fourquet and colleagues estimated the sickle-cell trait frequency to be 8.2% (Fourquet et al. 1974). The Buettner-Januschs also examined other serogenetic markers such as haptoglobin and transferrin types, their results being consistent with a hypothesis of African admixture in the ancestral Malagasy population (Buettner-Janusch and Buettner-Janusch 1964: 165–6). Significant differences in physical appearance between the highland Merina group and lowland/coastal non-Merina groups were observed by Singer and colleagues (Singer et al. 1957: 102–3, 114–5) and re-affirmed by the Buettner-Januschs (Buettner-Janusch and Buettner-Janusch 1964: 165–6) who also suggested that the Merina and the other plateau groups had an origin different from that of the coastal peoples—a hypothesis which implied more than one migration to the island—and that the original inhabitants settled on the plateau and the dominant groups (Merina) drove their neighbours to the coastal regions where more admixture with people from Africa occurred (Buettner-Janusch and Buettner-Janusch 1964: 168). However, the Buettner-Januschs were mindful that the low level of resolution of their genetic studies and the lack of comparative data from other groups, such as the Arabs, Indians and Europeans, made it impossible to accurately estimate their contributions to the gene pool of the islanders (Buettner-Janusch and Buettner-Janusch 1964: 168).

## *Beta-Globin Haplotype Analysis*

Colleagues in Jenkins' laboratory further investigated the distribution of the sickle-cell trait adopting a DNA approach that, through restriction fragment length polymorphisms (RFLPs), examined genetic variation within the beta-globin gene. This method facilitated an examination of DNA haplotypes (combinations of alleles found at different regions of the beta-globin gene) in people with sickle-cell anemia, enabling them to trace the origin of this trait among the Malagasy (Hewitt et al. 1996; Soodyall et al. 1996a). Using a much larger sample of individuals (1.425 individuals), the sickle cell trait ($\beta^S$) was found throughout Madagascar but at higher frequencies in some coastal populations (9.9% in East-Coastal; 2.3% in Southwest; 6.5% in West, and 5.8% in the North) compared with the frequency of 1.2% in Central Highland populations. This observation was not unexpected given the selective advantage sickle-cell anemia confers against malaria.

The ($\beta^S$) mutation is thought to have arisen at least five times in humans because there exist five distinct ($\beta^S$)-associated haplotypes. Four of these are exclusive to Africa; the 'Bantu' haplotype—found in Bantu-speaking peoples; 'Senegal' haplotype—in Atlantic West Africa; 'Benin' haplotype—in west Central Africa; and 'Cameroon' haplotype—found among the Eton group in Cameroon. The fifth haplotype, found in individuals from eastern Saudi Arabia and India, is referred to as the 'Arab-Indian' haplotype (Pagnier et al. 1984; Kulozik et al. 1986; Lapouméroulie et al. 1992). Twenty-five of the 35 ($\beta^S$) haplotypes found in the Malagasy were conclusively 'Bantu'; seven haplotypes matched the 'Bantu' haplotype at five of the seven sites used to derive the haplotypes (the phase could not be fully resolved at one or two sites in the remaining haplotypes) and were most probably 'Bantu', one haplotype displayed the characteristic 'Senegal' haplotype, and two were rare or atypical haplotypes. The Bantu ($\beta^S$) mutation is thought to have been introduced into Madagascar by Bantu-speaking immigrants (colonists or slaves) from Central or East Africa. The Senegal ($\beta^S$) mutation is likely to have been introduced by one or more Senegalese soldiers who were brought to Madagascar by the French, and who relieved the malaria-stricken French expeditionary force of 1894. The French expedition, was tasked with taking over the governing of the island from the Merina, but had been decimated by malaria—of the 6,725 fatalities sustained by the French expedition, only 25 were due to wounds sustained in battle, the remainder were caused by malaria (72 percent) and typhoid (12 percent) (Campbell 2005a: 323, 339). Whether the $\beta^S$ Senegal mutation helped protect the Senegalese soldiers from malaria, we shall never know. At least one of them passed on his Senegal ($\beta^S$) mutation—as evidenced by its presence in Madagascar at the present time (Hewitt et al. 1996: 1306–7).

The β-globin haplotype study by Trefor Jenkins and colleagues provided the first definitive biological evidence that a major component of Malagasy ancestry is derived from African populations, particularly Bantu-speaking Negroids. Betaglobin A haplotypes, consistent with the claim for a significant African contribution, are also suggestive of an Asian/Oceanic and Caucasoid admixture within the Malagasy.

### *Mitochondrial DNA Studies*

In the late 1980s, Rebecca Cann, Mark Stoneking and the late Alan Wilson demonstrated the value of mtDNA in reconstructing human origins (Cann et al. 1987), and subsequent mtDNA studies have continued to advance our understanding of historical human migration routes and assessing population affinities (Behar et al. 2008; Soodyall et al. 2008; Schlebusch et al. 2009). More specifically, mtDNA harbors information on the transmission of mtDNA lineages exclusively along maternal lines in a population over time. Since mtDNA does not undergo recombination, any changes are due to mutations only. The relationship of the divergent lineages can be deduced by phylogenetic analysis and linked to a MRCA discussed above; such studies have shown that the MRCA for all contemporary mtDNA lineages has an origin in sub-Saharan Africa dating to about 150,000–200,000 years ago (Behar et al. 2008: 1130, 1140).

mtDNA reveals how females have contributed to the gene pool, so a survey of mtDNA variation in Madagascar would elucidate the geographic regions of origins of females (the founding mothers) who have contributed to the contemporary Malagasy gene pool. If we assume a date of ~ 2300 years since the first presence of humans on the island, and if we assume that there were women among these ancestors, then the mtDNA history on the island spans 92 generations (using 25 years as a generation time). Based on the historical information of Madagascar, female ancestors of the contemporary population could have been introduced to the island during this time period from a broad geographic region. The value of using mtDNA in historical reconstruction is that the current global distribution of haplogroups (i.e. 'branches') of the human mtDNA tree is strongly correlated with geography. By examining certain regions of the mtDNA, such as the control region (sometimes also called the D-loop) using RFLPs or sequencing the whole mtDNA genome, it is possible to assign an individual's mtDNA to a haplogroup and infer its ancestry to a geographic region.

Another useful marker for population genetic studies is located in the region between the genes that code for transfer RNA for lysine ($tRNA^{Lys}$) and cytochrome oxidase II (COII). In most humans we find two copies of a segment of DNA made up of nine nucleotide base pairs (9-bp). Some people, however, have a deletion of one of these repeats, and when this happens we say that the individual has the 9-bp deletion trait to distinguish them from other individuals with two copies. Sometimes, some individuals have three copies of this 9-bp repeat. In our survey of 280 Malagasy individuals, we found that the 9-bp deletion was present at a frequency of 26.8% (see map 1) (Soodyall et al. 1995). Previous studies showed that this deletion occurred in contemporary and ancient Polynesians, and to a lesser extent in Indonesian and other Asian populations, and that it had also been found in Africa. Sequencing data, however, revealed that the 9-bp deletion arose more than once, at least once in Asia and again in Africa from independent events (Soodyall et al. 1996b). After sequencing the control region of the mtDNA molecule in Malagasy individuals who had the 9-bp deletion, we found that 70.7% of the mtDNA lineages could be traced to Asian origins and the remaining 29.3% to African origins. More importantly, sequencing data helped resolve the evolutionary sequence of events associated with the Asian form of the 9-bp deletion; those individuals who harboured the three nucleotide substitutions at positions 16217 (T→C), 16247 (A→C) and 16261 (C→T)—common among people from the Polynesia area—were said to have the "Polynesian motif." We found that 96.2% of Malagasy individuals with the Asian-form of the 9-bp deletion had the "Polynesian motif" and the data could not indicate if carriers of this motif came to Madagascar directly from Polynesia or from a common ancestral population in Indonesia.

Matthew Hurles and colleagues working on a smaller sample of Malagasy (37 individuals) confirmed our observations (Hurles et al. 2005). They found that 37.8% of the mtDNA in their sample originated in Africa. Of the remaining 62.2% that they traced to an Asian source, 73.9% could be assigned to haplogroup M (the commonest mtDNA haplogroup found in India) and 26.1% to haplogroup R. Using a statistical method, they performed a "centre of gravity" analysis to locate the most likely geographic region of origin of the mtDNA lineages in the Malagasy. Using published data in the comparative data, they found exact matches for all African lineages, but for only three of eight Asia-derived lineages. This could be due to the lack of sequence data available from Southeast Asian populations at the time.

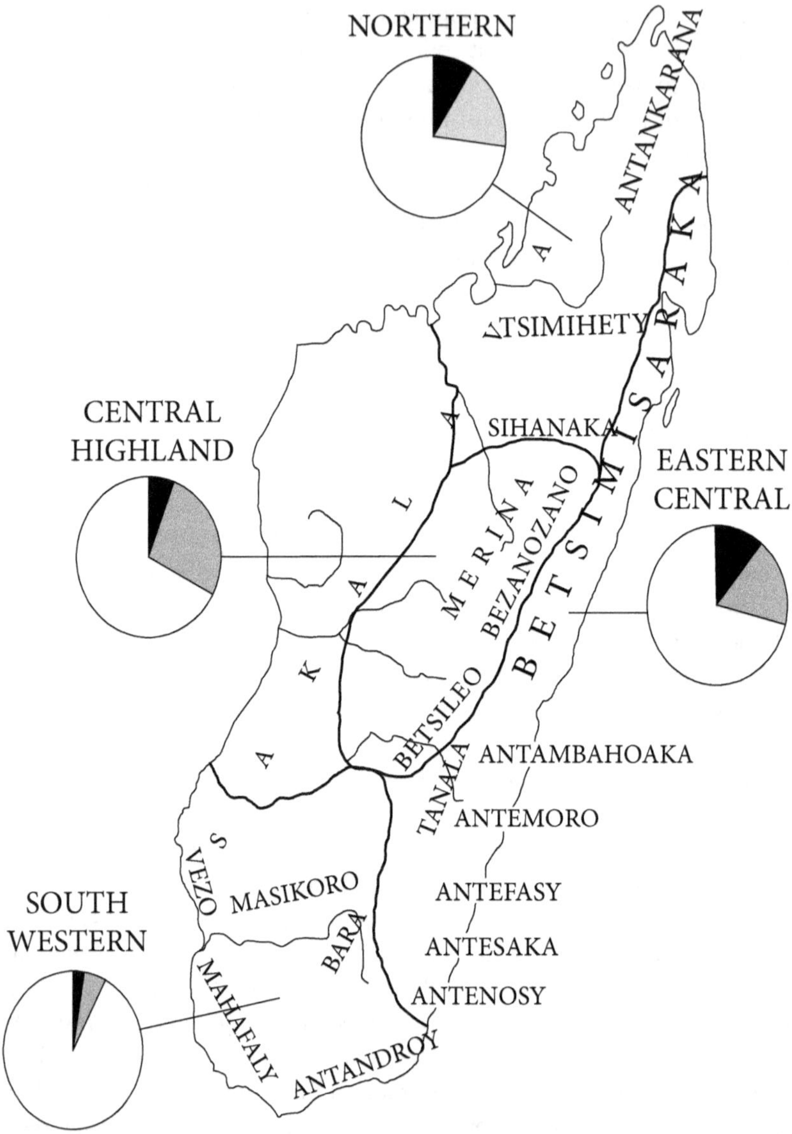


Map 1. Map of Madagascar showing the approximate location of ethnic groups and the frequencies of the 9-bp deletion (combined black and grey of the pie chart) and the relative contribution of the African (black) and Asian (grey) sources of the deletion in the Northern, Central Highland, South Western and Eastern Central regions.

More recently, Harilanto Razafindrazaka's group investigated the origin of the "Polynesian motif" in 266 individuals from three Malagasy ethnic groups (Merina, Vezo and Mikea). They suggested that the term "Polynesian motif" be replaced by the nomenclature used to assign the mtDNA sequences into haplogroups using their placement on a phylogenetic tree (a tree that shows the evolutionary relationship of the different mtDNA sub-branches). Using this nomenclature, sequences with the 9-bp deletion and the three substitutions mentioned earlier were placed on branch B4a1a1a (Razafindrazaka et al. 2010). In this sample, the overall frequency of the 9-bp deletion was 21.8%, with the highest frequency found among the Merina (50.0%) followed by the Vezo (21.8%) and Mikea (13.4%). This was in keeping with previous findings (Soodyall et.al. 1995: 378). Using mtDNA sequencing, Razafindrazaka and colleagues also confirmed the low diversity of mtDNA types associated within branch B4a1a1a in the Malagasy. Altogether, they resolved only four haplotypes in the sample; one formed the root and was common to all three groups occurring at frequencies of 44.7% in the Merina, 17.8% in the Vezo and 13.4% in the Mikea. The remaining three haplotypes differed from the common haplotype at one nucleotide position and was specific to a single ethnic group (i.e. either in the Merina, Vezo or Mikea).

Razafindrazaka's team also sequenced the complete mtDNA genomes of three individuals (one from each of the major haplotypes associated with B4a1a1a) to assess how well the three substitutions found in the control region were correlated with substitutions elsewhere on the mtDNA genome (note: whole mtDNA sequencing has been used widely to more accurately reconstruct the mtDNA phylogeny previously suggested on the basis of only the control region sequence and a few informative RFLPs). They used these data, in conjunction with published data from an additional 15 individuals, to conduct phylogenetic analysis. Two new substitutions at nucleotide positions 1473 and 3423 were identified in all three Malagasy sequences. A separate survey for these polymorphisms confirmed their presence in 58 Malagasy with the "Polynesian motif," but they were not found in Melanesian, Polynesian and Island Southeast Asian individuals for who complete mtDNA genome sequences were available. This prompted the authors to coin the term "Malagasy motif" which incorporated the presence of these two new mutations in association with the B4a1a1a sequences.

It is worthwhile reviewing here the evolutionary history of the "Polynesian motif" in order to better contextualise the origin of the "Malagasy motif." The "Polynesian motif" gained popularity given its high frequency among Polynesians. Using more recent mtDNA whole genome sequence data, the polymorphisms that define the subhaplogroup B4a1a1a are the substitutions at positions 14022 (A→G) (a new mutation associated with the "Polynesian motif" found outside the control region), and the three substitutions—16217 (T→C), 16247 (A→G) and 16261 (C→T)—that occur within the control region. It has been suggested that this lineage probably evolved in eastern Island Southeast Asia or Near Oceania around 6,200–10,900 years ago (Pierson et al. 2006). The precursor to this haplogroup that lacks the 16247 (A→G) substitution has been found in Taiwanese aboriginal groups with an estimated age of 13,200 years (Trejaut et al. 2005). These dates are consistent with a model whereby Austronesian-speaking populations expanded out of Taiwan during the mid- to late-Holocene to cover a vast area from Taiwan in the north, to New Zealand in the south, and from remote Polynesia in the east to Madagascar in the far west (Razafindrazaka et. al. 2010: 578). The "Polynesian motif" is currently found at highest frequency in Polynesia, reaching fixation (100%) in some populations. It is also common in Micronesia and parts of Near Oceania where it is not necessarily restricted to Austronesian-speaking populations. The "Malagasy motif" was dated to 6,000 years before present (95% confidence 1,200–16,800 YBP) and is found exclusively in the Malagasy. Given the limitations of molecular dating and that demographic expansions (related to DNA sequences) may have predated geographic expansions (related to populations), the authors acknowledge that these dates are only broadly consistent with archaeological and paleoecological estimates for the first settlement of Madagascar around 2,300 years ago. Its absence in Melanesian and Polynesian individuals used in the comparative analysis, suggests that the "Malagasy motif" expanded in a relatively short time after its appearance in Madagascar. Subsequently, the motif accumulated distinct mutations in each Malagasy population, which is consistent with long-standing isolation of the groups from each other.

More recently, Murray Cox and colleagues attempted to localize where the "Malagasy motif" originated (Cox et al. 2012). They did this by screening for the "Polynesian motif" in 2745 individuals sampled from 12 island groups spanning the Indonesian archipelago (Sumatra, Nias, Mentawai, Java, Bali, Sulawesi, Sumba, Flores, Lembata, Alor, Pantar and Timor). The "Polynesian motif" was only found in 45 individuals (1.6%) but restricted to four islands [Bali (1.2%), Sumba (0.6%), Alor (4.4%), and Timor (6.4%)]. None of these individuals carried the diagnostic markers for the "Malagasy motif" described earlier. Moreover, their maximum-likelihood estimates suggest that Madagascar was settled approximately 1200 years ago by a very small group of around 30 women mostly (93%) from Indonesia (Cox et al. 2012).

Sergio Tofanelli and colleagues developed a statistical method to assess population affinities based on haplotype sharing ($D_{HS}$). This method is less prone to sample size and useful in a time period spanning 200 generations (this is equivalent to 5,000 years if a generation time of 25 years is used and 4,000 years when using a generation time of 20 years) (Tofanelli et al. 2009). They performed multilevel analysis in which the diversity among the main Malagasy ethnic groups could be assessed, including refinements of lineage ancestries, and evaluating models of inheritance concurrently. Their findings revealed that: (1) the main source of genetic variation among the islanders separated the highland groups from the coastal groups; (2) the Indonesian (Asian) to African contributions to the mtDNA pool was similar for the highland (63%:37%) and coastal (62%:38%) groups; (3) the B4a1a1 sequences accounted for 34.3% of mtDNA lineages; Malagasy Indonesian-type sequences matched closely to Insular Southeast Asian haplotypes, in particular samples from Molucca (Ambon; $D_{HS}$ = 0.532) and Sunda (Sulawesi, Lombok and Borneo; $D_{HS}$ 0.544, 0.566, 0.585, respectively) island groups. [Note $D_{HS}$ values closer to 0 are indicative of more sharing compared with 1 which suggests no sharing].

These authors also presented for the first time some indication of a founding female population size and time of arrival in Madagascar. The African ancestry was estimated to be 1,820–3,820 years ago for the coastal groups from an effective population size ($N_e m$) of 1,000–2,000 founding females (effective population size refers to the breeding population). The Asian contribution was estimated to have occurred over 3,500 years ago with an effective population size of 200 females for the highland group. Again, the reasons for the inconsistency with the molecular dating estimates and the dates for settlement in Madagascar, suggested on the basis of archaeological data apply to this estimate.

Vincent Dubut and co-workers explored the connection between India and Madagascar and sequenced three mtDNA genomes of Malagasy individuals whose mtDNA haplogroups were assigned to haplogroups M26 and M23 (Dubut et al. 2009). They suggested that the presence of haplogroup M23 in Madagascar provides a link between India and Madagascar suggesting that the migration from Indonesia to Madagascar involved a 'stop-over' in India. However, independent migrations from India which have also occurred in more recent times would, presumably, have contributed to the Malagasy gene pool. This finding underscores the genetic complexity of the Malagasy population and ought to be explored further in studies with larger samples.

While mtDNA has been an informative tool in tracing the maternal history with respect to African and non-African origins of the present-day Malagasy, it has not delivered on resolving questions related to the timing of settlement. Different dates for settlement have been suggested (ranging from 1,200 years – 3,800 years). These inconsistencies may be partly due to differences in estimates of the mutation rate for mtDNA and the larger standard errors around these estimates.

### *Y Chromosome DNA Studies*

The Y chromosome contains the largest non-recombining block in the human genome and is therefore extremely important for evolutionary genetic studies. While studies on mtDNA describe the maternal history of a population, the paternal history can be inferred from using the Y chromosome. The first Y chromosome polymorphism was reported in 1985 (Casanova et al. 1985) but more than a decade elapsed before a well-resolved Y chromosome tree was available (Hammer et al. 2001; Underhill et al. 2001; YCC 2002). The structure of this tree is primarily based on binary polymorphisms, and specific branches are assigned to haplogroups following a hierarchical pattern. Currently the Y chromosome tree consists of 20 major clades containing 311 distinct haplogroups defined by 599 mutational events (Karafet et al. 2008). The addition of Y chromosome short tandem repeat (STR) data to haplogroup information further resolved the Y chromosomes into haplotypes (Underhill and Kivisild 2007).

We had originally used a low resolution typing system to examine Y chromosome variation in the Malagasy (Morar 2000: 157–211). Later, we introduced 22 additional biallelic markers and eight STRs to increase the resolution in 439 males. Based on the bi-allelic markers, 16 haplogroups were identified. Admixture analysis indicated that the contributions to the Malagasy Y chromosome pool from sub-Saharan Africa, East Asia and Eurasia was 68%, 15% and 15%, respectively. There was no detectable male-mediated gene-flow from Oceania. The analysis of STR variation within the major haplogroups provides evidence for bottleneck and founder effects that accompanied the settlement events (Soodyall, unpublished).

Hurle's group also examined Y chromosome variation among Southeast Asians and the Malagasy using a higher resolution of analysis based on

44 polymorphisms in 434 males (35 Malagasy) (Hurles et al. 2005: 895–6). Using the differences in the global geographic distribution of Y chromosome haplogroups, they were able to assign 51.4% of the Y lineages to African origins, and the remaining 48.6% to non-African origins—of which 34% were from Southeast Asia or Oceania (haplogroup O). Acknowledging that the sample size is small, the authors performed a statistical test (pairwise $F_{ST}$ distances implemented in Arlequin software) to determine the closest populations in terms of genetic distance to the Malagasy. Among the populations they examined as potential ancestral populations from Island Southeast Asia and Oceania (Banjarmasin and Kota Kinabalu from Borneo; Taiwan; Philippines; Papua New Guinea; Vanuatu, Cook Islands, W. Samoa, Majuro and Kapingamarangi) including two East African samples from Kenya and Tanzania from published data, they identified the two populations from Borneo as the best candidates for the likely source of these lineages. This observation was based on the presence of appreciable frequencies of lineages O1b and O2a* in both populations as well as a relative lack of chromosomes belonging to O3 lineages. The closest single Island Southeast Asian or Oceanic population to the Malagasy is the Banjarmasin (Hurles et.al. 2005: 900).

They also used three different methods to estimate the proportion of African ancestry in Malagasy paternal lineages; all three methods yielded estimates ranging between 56–58%, which is similar to the results from lineage counting. Given that the common Y chromosome haplogroups found in Polynesia (O3 and C) were not found in this small sample, they suggested that the mtDNA "Polynesian motif" could not have come directly from a Polynesian population.

Using a larger sample of males (N = 133), Tofanelli's group estimated the proportion of Indonesian to African ancestry to be 39%:50% for the highland groups and 20%:74% in the coastal populations (Tofanelli et.al. 2009: 2114). Recent contributions from Eurasia, the Indian subcontinent and the horn of Africa-Arabia were estimated to be 11% in the highland group and 4% in the coastal groups. Regardless of the unequal apportionment, Asian and African components were defined by the same haplogroups in the two ethnic subgroups. The male African component is compatible with a large size range of founders (200–2,000) spanning a period between c.1200–1935. They could not perform simulations to obtain these estimates for the Asian ancestry due to lack of comparative data from the presumed parental populations. Thus, in contrast to mtDNA studies, Y chromosome data are consistent with a larger African input to the gene pool of the Malagasy.

To overcome some of the sex-based biases when assessing the origins of parental populations in gene pools, the next section examines how autosomal DNA markers – those markers which both parents contribute equally to the offspring, have contributed to our understanding of the genetic heritage of the Malagasy.

### *Autosomal DNA Studies*

Autosomal DNA markers carry information inherited from both parents on 22 pairs of chromosomes (1 to 22). Together with the sex chromosomes (XX in females and XY in males), they make up the 23 pairs of chromosomes (46 in total) found in humans. The human genome consists of roughly 3 billion bp of DNA and is made up of coding (genes) and non-coding regions. With the exception of identical twins, all humans differ from each other, with an average of one difference being found in about 1000 bp of DNA in the human genome.

In addition to mtDNA and Y chromosome DNA markers, stretches of repeat DNA—long (LINES) and short tandem repeats (STRs), Alu insertions, etc., have been used to study genetic variation in the human genome. Because of their high variability and ease of identification, STRs are extremely valuable for mapping genes and in the diagnosis of inherited disease, in forensic and parentage testing and for inferring phylogenetic relationships among human populations. In this section we discuss the value of four autosomal STRs [*HUMCSF1PO* in the human c-*fms* proto-oncogene for the CSF-1 gene; *HUMTHO1* in the human tyrosine hydroxylase gene; *HUMCD4* within the CD4 gene, and an $(AC)_n$ repeat in the dopamine receptor D2 (*DRD2*) gene] to reconstruct the history of the Malagasy because they show distinct allele frequency distributions in African and non-African populations) (Hammond et al. 1994; Kidd et al. 1998).

The Malagasy sample included two highland [Merina (MR) and Betsileo (BT)] and nine coastal groups [Antaisaka (AS), Tsimihety (TS), Sakalava (SK), Mahafaly (MF), Antandroy (AD), Betsimisiraka (BS), Antaimoro (AM), Antankarana (AK) and Mikea (MK)]; 11 of the 18 ethnic groups on the island. The non-African sample consisted of Indonesians [South Sulawesi (Ujung Pandang), Banjamarsin (southern Kalimantan), Alor, Medan, Sumba, Bangka, Ambon and Palembang], Melanesians [Bougainville (Nasioi) in the Solomon Islands and Gimi and Goroka in Papua New Guinea], Polynesians (Tonga and Samoa), and a sample of South African Indians. The African sample included Zambians, North Sotho (Pedi) and Tsonga from South Africa and the Beya (Gbaya, Mbimou and Sangha Sangha) and Mbenzele from the Central African Republic (CAR). These 'putative' parental populations have been sampled from geographic areas in Africa and Asia most likely to have contributed to the Malagasy gene pool and for which samples were available at the time the research was conducted (Morar 2000: 10–36).

Allele frequency distributions in the Malagasy at the four loci are discussed in the context of variation in highland and lowland groups as compared to African and non African populations (Morar 2000: 78–99). At all four loci, the Malagasy groups are relatively homogeneous and show varying degrees of similarities to both the Indonesian and African populations. The data show complex trends, with Merina often differing significantly from the Betsileo (a highland group) as well as from coastal groups. On average, however, the two highland groups showed less affinity with the Tsonga and Zambian populations from Africa than they did with the Malagasy coastal populations.

We also examined the genetic affinities between the Malagasy and the African and Asian populations used in the comparative analysis (see figure 1). The genetic affinities between Malagasy groups are stronger (inferred from the smaller values for genetic distances between them [0.000 to 0.099]) than those between the Malagasy and other populations (African and Asian groups [0.001 to 0.249)]. In addition, there was a greater affinity between Malagasy highlanders and Indonesians, and between coastal Malagasy and African populations (Morar 2000: 101–2). These findings are consistent with other genetic studies discussed above.

The STR frequency data were also used to estimate the proportions of admixture in the highland and lowland Malagasy populations from two parental populations. Two assumptions were made for the analysis: (1) admixture affects all loci equally and simultaneously, whether gene flow occurs in large amounts over a few generations or in small amounts over many generations, and (2) the allele frequencies in the parental populations and the 'hybrid' population have not drifted since the time when admixture occurred. Pooled data from the Tsonga and Zambians represented the African parental population and the Indonesians [south Sulawesi (Ujung Pandang), Banjamarsin (southern Kalimantan), Alor, Medan, Sumba, Bangka, Ambon and Palembang] the Asian parent population. The highland and coastal Malagasy populations represented the hybrid populations in the analysis. Using allele frequency data from the four loci, the African contribution was estimated to be 32.5% in the highlanders and 58.8% in the coastal groups (Morar 2000: 114).

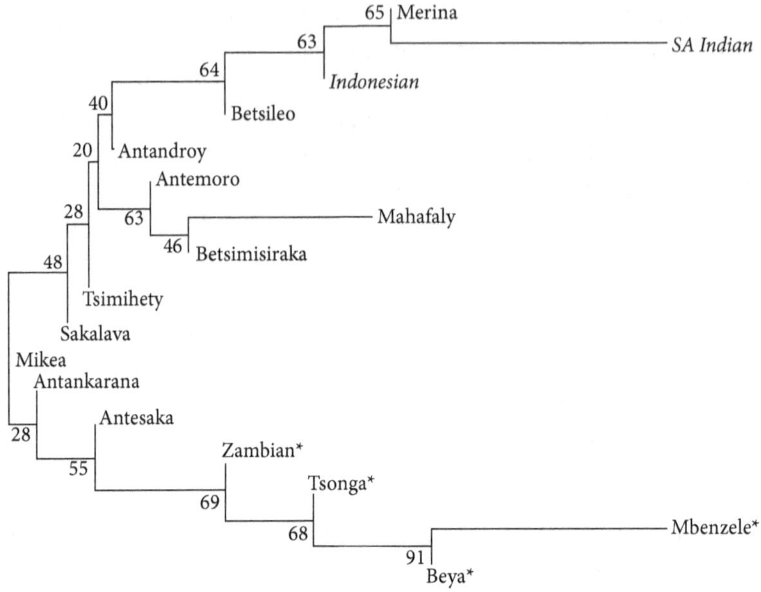


Figure 1. Phylogenetic tree showing the affinities of the Malagasy with some African (population names succeeded with *) and Asian (in italics) populations. The choice of populations used in the comparative analysis was based on availability of samples that were examined at the same level of resolution and does not necessarily suggest that these are the parental groups who contributed to the Malagasy gene pool. However, these data support the view that the highland groups (Merina and Betsileo) have a stronger affinity with Asian populations compared with the other Malagasy groups.

## *Conclusion*

Reconstructing the prehistory of the Malagasy has been the subject of numerous studies by historians, anthropologists, archaeologists, linguists, paleobotanists and geneticists. Campbell has provided a concise review on the subject and highlighted how each discipline has contributed in shedding more light on the subject and also on the debates that still prevail among the research communities (Campbell 2005b). Having reviewed the different kinds of genetic studies that have been conducted to date, it is apparent that genetic data supports the view of African and Indonesian admixture suggested on the basis of historical, linguistic and archaeological data. However, genetic data cannot resolve where and how the admixture took place. Did admixture occur prior to colonization or following colonization? If contact between Indonesian traders and Bantu-speakers in East Africa resulted in gene admixture between the two groups, the proto-Malagasy would have assimilated African genes into their gene pool. The fact that the present-day Malagasy way of life and language still reflects a dominance of Indonesian culture indicates that these two traits were not affected substantially by any time the Indonesians might have spent in Africa. The coherence and strength of a common language and culture throughout the island also supports the idea that new immigrants to the island from different cultures (Jews, Persians, Indians and even Chinese) had minor effects on the lifestyle of the settlers. Only minor isolated cultural and vocabulary links are seen except for techniques relating to cattle herding which were acquired from Africa (Brown 1978: 10–15, 17–29; Mack 1986: 14–15).

It is also possible that admixture took place in the island following initial human colonization. If it is assumed that Madagascar was first colonized by Indonesians, the African ancestry demonstrated in the present study could be due to either African slave imports or/and voluntary African migration to Madagascar. Over time, there is little doubt that the slave trade influenced the Malagasy gene pool across the island because slaves freed at the end of the nineteenth century were assimilated into the general Malagasy settlements where they adopted the Malagasy culture. The close relationship between Malagasy groups revealed in the present study is compatible with this widespread assimilation. Admixture calculations support previous data of a larger African contribution to lowland groups (58.80%) than to the highland groups (32.45%). A problem that may be biasing this estimate is that people can and do chose their own ethnic identity especially when they are not living near their place of birth. Because of the stigma attached to being a slave, people of slave ancestry generally claim to be members of the same ethnic groups as their neighbours and do not refer to their families' actual origins.

It is difficult to determine the level of variation that existed within the founding Malagasy population/s at the time/s of settlement. The fluctuations in the Malagasy gene pool during the last century make it extremely difficult to establish the size or composition of the founding population of Madagascar. Also, it is difficult to distinguish genetic contributions made by founders, whether from Africa, SE Asia or elsewhere, who contributed to the gene pool of the Malagasy as a result of different waves of migration. It would be desirable for future research to adopt a multidisciplinary approach that exploits the use of genetic data in conjunction with historical, anthropological, archaeological, paleoecological and climatic data, to shed more light on the origins of the people of Madagascar.

## *Bibliography*


Behar, D.M., Villems, R., Soodyall, H., Blue-Smith, J., Pereira, L., Metspalu, E., Scozzari, R., Makkan, H., Tzur, S., Comas, D., Bertranpetit, J., Quintana-Murci, L., Tyler-Smith, C., Wells, R.S., and Rosset, S; Genographic Consortium. 2008. "The dawn of human mitochondrial DNA diversity." *American Journal of Human Genetics* 82: 1130–1140.

Brown, M. 1978. *Madagascar Rediscovered.* London: Damien Tunnacliffe.

Burney, D.A. 2005. "Finding the connections between paleoecology, ethnobotany, and conservation in Madagascar." *Ethnobotany Research & Applications* 3: 385–389.

Burney, D.A. 1987. "Late Holocene vegetational change in central Madagascar." *Quaternary Research* 28. 1: 130–143.

Burney, D.A., Burney, L.P., Godfrey, L.R., Jungers, W.L., Goodman, S.M., Wright, H.T., and Jull, A.J.T. 2004. "A chronology for late prehistoric Madagascar." *Journal of Human Evolution* 47: 25–63.

Buettner-Janusch, J., and Buettner-Janusch, V. 1964. "Hemoglobins, haptoglobins, and transferrins in the peoples of Madagascar." *American Journal of Physical Anthropology* 22: 163–169.

Campbell, G. 2005a. *An Economic History of Imperial Madagascar, 1750–1895: The Rise and Fall of an Island Empire.* Cambridge: Cambridge University Press.

Campbell, G. 2005b. "The Debate over Malagasy origins" *ZIFF Journal* 2: 5–14.

Cann, R.L., Stoneking, M., and Wilson, A.C. 1987. "Mitochondrial DNA and human evolution." *Nature* 325: 31–36.

Casanova, M., Leroy, P., Boucekkine, C., Weissenbach, J., Bishop, C., Fellous, M., Purrello, Fiori, M.G., and Siniscalco, M. 1985. "A human Y-linked DNA polymorphism and its potential for estimating genetic and evolutionary distance" *Science* 230: 1403–1406.

Cox, M.P., Nelson, G.N., Tumonggor, M.K., Ricaut, F-X., and Sudoyo, H. 2012. "A small cohort of Island Southeast Asian wowen fouded Madagascar." *Proceedings of the Royal Society B* 279: 2761–2768.

Dubut, V., Cartault, F., Payet, C., Thionville, M-D., and Murail, P. 2009. "Complete mitochondrial sequences for haplogroups M23 and M46: insights into the Asian ancestry of the Malagasy population." *Human Biology* 81: 495–500.

Fourquet, R., Sarthou, J., Roux, J., and Aori, K. 1974. "Hémoglobine S et origines du peuplement de Madagascar: Nouvelle hypothese sur son introduction en Afrique." *Archives Institute Pasteur Madagascar* 43: 185–220.

Hammer, M.F., Karafet, T.M., Redd, A.J., Jarjanazi, H., Santachiara-Benerecetti, S., Soodyall, H., and Zegura, S.L. 2001. "Hierarchical patterns of global human Y-chromosome diversity" *Molecular Biology and Evolution* 18: 1189–1203.

Hammond, H.A., Jin, L., Zhong, Y., Caskey, C.T., and Chakraborti, R. 1994. "Evaluation of 13 short tandem repeat loci for use in personal identification applications" *American Journal of Human Genetics* 55: 175–189.

Hewitt, R., Krause, A., Goldman, A., Campbell, G., and Jenkins, T. 1996. "B-Globin haplotype analysis suggests that a major source of Malagasy ancestry is derived from Bantu-speaking Negroids" *American Journal of Human* Genetics 58: 1303–1308.

Humbert, H. 1927. "Destruction d'une flore insulaire par le feu." *Mémoires de l'Académie Malgache* 5: 1–80.

Hurles, M.E., Sykes, B.C., Jobling, M., and Forster, P. 2005. "The dual origin of the Malagasy in Island Southeast Asia and East Africa: Evidence from maternal and paternal lineages" *American Journal of Human Genetics* 76: 894–901.

Karafet, T.M., Mendez, F.L., Meilerman, M.B., Underhill, P.A., Zegura, S.L., and Hammer, M. 2008. "New binary polymorphisms reshape and increase resolution of the human Y chromosomal haplogroup tree." *Genome Research* 18: 830–838.

Kidd, K.K., Morar, B., Castiglione, C.M., Zhao, H., Pakstis, A.J., Speed, W.C., Bonne-Tamir, B., Lu, R.B., Goldman, D., Lee, C., Nam, Y.S., Grandy, D.K., Jenkins, T., and Kidd, J.R. 1998. "A global survey of haplotype frequencies and linkage disequilibrium at the DRD2 locus" *Human Genetics* 103: 211–227.

Kulozik, A.E., Wainscoat, J.S., Serjeant, G.R., Kat, B.C., Al-Awamy, B., Essan, G.J.F., Falusi, A.G.S., Haque, K., Hilali, A.M., Kate, S., Ranasinghe, W.A.E.P., and Weatherall, D.J. 1986. "Geographical survey of beta S-globin gene haplotypes: evidence for an independent Asian origin of the sickle-cell mutation." *American Journal of Human Genetics* 39: 239–244.

Lapouméroulie, C., Dunda, O., Ducrocq, R., Trabuchet, G., Mony-Lobé, M., Bodo, J.M., Carnevale, P., Labie, D., Elion, J., and Krishnamoorthy, R. 1992. "A novel sickle cell mutation of yet another origin in Africa: the Cameroon type." *Human Genetics* 89: 333–337.

Leopold, A. 1987. *A Sound County Almanac*. Oxford University Press.

Mack, John. 1986. *Madagascar, Island of the Ancestors*. London: British Museum.

Morar, B. 2000. "An exploration of the origins of the Malagasy using genetic polymorphisms" PhD Thesis, University of the Witwatersrand.

Pagnier, J., Mears, J., Dunda-Belkhodja, O., Schaefer-Rego, K., Beldjord, C., Nagel, R., and Labie, D. 1984. "Evidence for the multicentric origin of the sickle cell haemoglobin gene in Africa" *Proceedings of the National Academy of Science, USA* 81: 1771–1773.

Pierson, M.J., Martinez-Arias, R., Holland, B.R., Gemmell, N.J., Hurles, M.E., and Penny, D. 2006. "Deciphering past human population movements in Oceania: probably optimal trees of 127 mtDNA genomes." *Molecular Biology and Evolution* 23: 1966–1975.

Razafindrazaka, H., Ricaut, F.X., Cox, M.P., Mormina, M., Dugoujon, J.M., Randriamarolaza, L.P., Guitard, E., Tonasso, L., Ludes, B., and Crubézy, E. 2010. "Complete mitochondrial DNA sequences provide new insights into the Polynesian motif and the peopling of Madagascar." *European Journal of Human Genetics* 18: 575–581.

Schlebusch, C.M., Naidoo, T., and Soodyall, H. 2009. "SNaPshot minisequencing to resolve mitochondrial macro-haplogroups found in Africa" *Electrophoresis* 30: 3657–3664.

Singer, R., Budtz-Olsen, O.E., Brain, P., and Saugrain, J. 1957. "Physical features, sickling and serology of the Malagasy of Madagascar." *American Journal of Physical Anthropology* 15: 91–124.

Soodyall, H., Jenkins, T., Hewitt, R., Krause, A., and Stoneking, M. 1996a. "The peopling of Madagascar." In Boyce, A.J. and Mascie-Taylor, C.G.N. Eds. *Molecular biology and human diversity*. Cambridge: Cambridge University Press: 156–170.

Soodyall, H., Vigilant, A., Hill, A.V., Stoneking, M., and Jenkins, T., 1996b. "mtDNA control-region sequence variation suggests multiple independent origins of an 'Asian-specific' 9-bp deletion in sub-Saharan Africans" *American Journal of Human Genetics* 58: 595–608.

Soodyall, H., Jenkins, T., and Stoneking, M. 1995. "'Polynesian' mtDNA in the Malagasy" *Nature Genetics* 10: 377–378.

Soodyall, H., Makkan, H., Haycock, P., and Naidoo, T. 2008. "The prehistory of the Khoe and San" *Southern African Humanities* 20: 37–48.

Tofanelli, S., Bertoncini, S., Castrì, L., Luiselli, D., Calafell, F., Donati, G., and Paoli, G. 2009. "On the origins and admixture of Malagasy: new evidence from high-resolution analyses of paternal and maternal lineages." *Molecular Biology and Evolution* 26: 2109–2124.

Trejaut, J.A., Kivisild, T., Loo, J.H., Lee, C.L., He, C.L., Hsu, C.J., Li, Z.Y., and Lin, M. 2005. "Traces of archaic mitochondrial lineages persist in Austronesian-speaking Formosan populations." *PLoS Biology* 3: 1362–1372.

Underhill, P.A., and Kivisild, T. 2007. "Use of Y chromosome and mitochondrial DNA population structure in tracing human migrations" *Annual Review of Genetics* 41: 539–564.

Underhill, P.A., Passarino, G., Lin, A.A., Shen, P., Mirazon Lahr, M., Foley, R.A., Oefner, P.J., and Cavalli-Sforza, L.L. 2001. "The phylogeography of Y chromosome binary haplotypes and the origins of modern human populations." *Annals of Human Genetics* 65: 43–62.
YCC. 2002. "A nomenclature system for the tree of human Y-chromosomal binary haplogroups" *Genome Research* 12: 339–348.

# DON'T WE ALL WANT A WORLD FILLED WITH "BRIGHT FACES" AND "FAT-CHEEKED BABIES"? CREATING THE STATE AND CRAFTING IDEOLOGY IN EIGHTEENTH CENTURY IMERINA

Victor Raharijaona and Susan Kus

Political conflict and persecution have perpetuated horrific human misfortune, prehistorically, historically and, tragically, into the present day. Yet, we are also reminded daily by images in the press, from New Orleans to China to Haiti, of the horrors of famine, of disease, and of the natural disasters of flood, fire, earthquake, tsunami, cyclone, tornado and hurricane. Health and well-being seem to be so absent from the contemporary human condition! Yet health and well-being are, one could argue, a cross-cultural universal preoccupation and a hope that springs eternal from the human condition. Understandably the themes of health and well-being have become tied to arguments used both to legitimate and to condemn political regimes and claimants to authority, both religious and secular. Contemporary tragic and troubling examples are the commentaries of (Reverend) Pat Robertson and Rush Limbaugh concerning the earthquake of January 12, 2010 in Haiti. Robertson blamed this horrific tragedy on a pact with the devil made earlier by Haitians to overthrow French colonial authority, and the right wing political pundit, Rush Limbaugh, blamed the tragedy on communism, further arguing that the American President, Barack Obama, would put the tragedy to political use "to burnish [his and the Democratic Party's] credibility with ... both [the] light-skinned and dark-skinned black community in this country" (CNN 2010; Rushlimbaugh 2010).

Madagascar has and continues to face recurrent problems of human suffering often resulting from a concatenation of environmental and political factors. In an article that appeared in the *Madagascar Tribune* online on January 18, 2010, the journalist Régis Michel spoke in an open letter to the "Monsieur le Pazenkor-Président," the current (or *pas encore*) detainer of political authority in Madagascar. He remarked that when recently he received a letter and emails from friends in Madagascar with pictures of their families, he was overwhelmed by the image of a newborn whose mother was exceedingly thin. In another photo he saw "sunken faces, new wrinkles, and strained smiles." He further remarked that previously his friends had been "healthy and confident" but now they were "very thin and sad."[1] The letter continued as a plea to the government to resolve the political crisis in Madagascar. What is noteworthy in this letter for the purposes of our discussion below is that, without intention on his part, Michel virtually recreates some of the descriptive vocabulary and images of social and physical suffering that were used in the royal discourse of the eighteenth century attributed to Andrianampoinimerina. Andrianampoinimerina claimed to restore unity and peace to Imerina in order to fill the countryside with fat-cheeked babies, well-fed citizens, and bright smiling faces that were smooth and without wrinkles. (See Figure 1.)

Under the rule of Andrianampoinimerina (1787–1810), Imerina constituted an independent polity in the central highlands of the island. At its inception this state occupied an area of approximately 30 kilometers radius around the present day capital of Antananarivo. Under Andrianampoinimerina's son and immediate successor, Radama I, the Merina polity went on to bring one-third of this fourth largest island in the world under its control. The *Tantara ny Andriana eto Madagascar* (The History of the Sovereigns of Madagascar) (Callet 1981), a document of over 1,000 pages, offers an indigenous account of the history and politics of the Merina polity from its inception to its situation in the mid-to-late nineteenth century, when this document was recorded in the Merina dialect of the Malagasy language by the French Jesuit missionary, François Callet. A number of Callet's informants were descendants of Andrianampoinimerina's family and familiars, some of whom may even have witnessed the last days of Andrianampoinimerina's rule (Delivré 1974: 37–60). Understandably, the majority of the text recounts the deeds and words of this paramount sovereign. We draw the reader's attention to the fact that there is much evidence in these recorded oral traditions of a hegemonic political discourse that equates the health and well-being of Andrianampoinimerina's charges to his legitimate positioning as singular sovereign of Imerina, and additionally that this discourse of health and well-being, of "bright faces" and "fat-cheeked babies" (cf. Callet 1981: 808–809) was used to craft and legitimate an expansionist political agenda to

[1] *"...l'image de nouveau-né dont la maman était bien maigre.... Sur ces images, tous avaient le visage creusé, des rides nouvelles et le sourire en berne.... Ceux que j'avais quittés bien portants et confiants un an plus tôt n'étaient plus les mêmes !"*

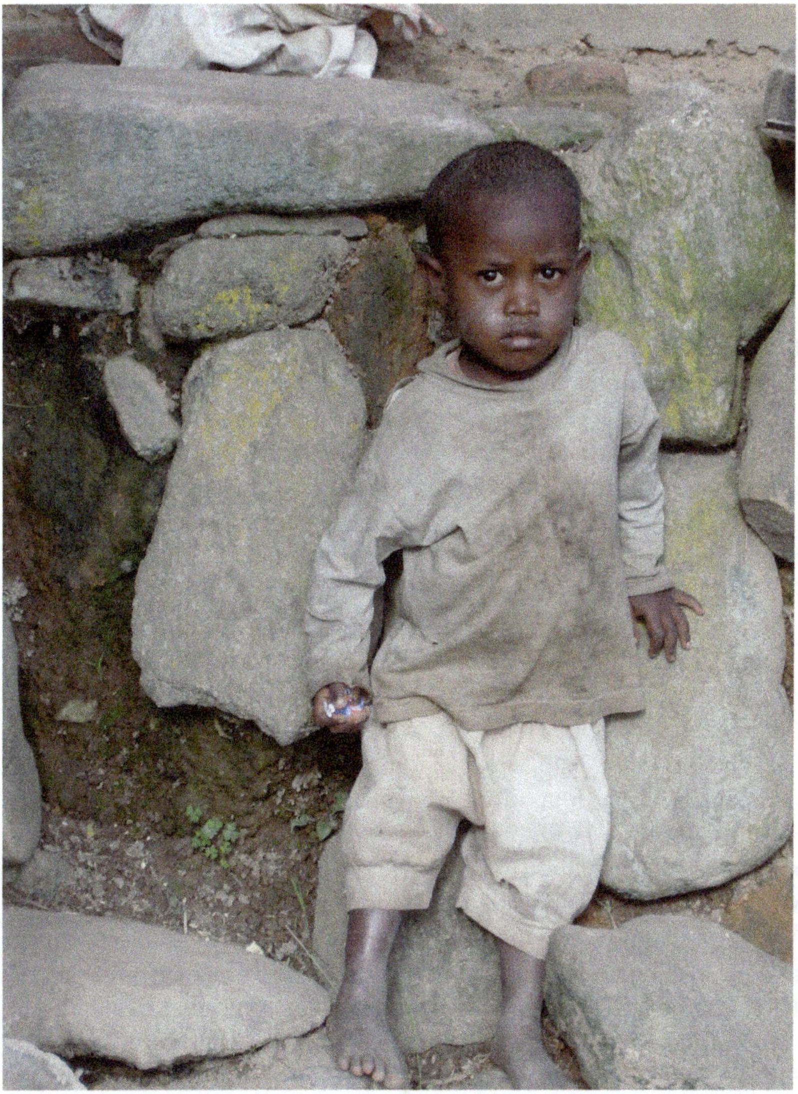

Figure 1. Young Malagasy in the Bestileo countryside, 2006.

bring the entire island under his rule. We propose to explore in more detail below the "state-crafting" of an ideology wherein the "the land and the polity" (*ny tany sy ny fanjakana*) were inextricably bound and wherein natural disasters and political enemies were confounded, further continuing our argument that the ideologies of early 'state' polities were neither gratuitous nor fortuitous (cf. Kus and Raharijaona 2000; 2006a; 2008).

## *A 'State' of Mind and a Matter of 'State'*

Anthropological archaeologists have contributed significantly to the discourse concerning the origins and functioning of complex polities, traditionally labeled states. In his McDonald lecture at Cambridge entitled "Process and Agency in Early State Formation" the archaeologist Kent Flannery "examines five agents who modified ideologies and created states from chiefdoms" (1999: 3). Included among them is Andrianampoinimerina. Flannery identifies "ecological, demographic and technological" (ibid.) processes that are "evident during state formation includ[ing such things as] chiefly cycling ... territorial expansion and the gaining of competitive advantage" (ibid.). Flannery paints a dramatic, if not 'epic,' storyline which each of his cases (i.e., Shaka, Osei Tutu, Mir Salim Khan, Kamehameha, Andrianampoinimerina) follows. These 'chiefs', these agents, came to power through audacity, testosterone, ruse, brute physical and military force, corvée labor, and territorial expansion. And then, almost as an afterthought, with impudent force and hegemonic flourish they co-opted ideology. Why? Because as Flannery suggests, the 'agentic' usurpers needed 'legitimacy'. Flannery says of the various usurpers: "While he himself is a product of process, the agent has a special role in state formation: he changes the ideology of his society to one more appropriate for a state" (1999: 15). Herein lays an interesting question: what is an ideology more appropriate for a state? According to Flannery, "[It is one that involves] breaking down the autonomy of subject peoples and redirecting their loyalty to the central figure of an emerging state." Flannery goes on to add: "To be sure, each ideological change is unique in terms of the symbolic repertoire of the culture involved" (1999: 15–17). In Flannery's effort at theoretical comprehensiveness, ideology is incorporated into the story of state origins, but it still is an epiphenomenon of history. The specifics of any ideology are 'unique' yet they appear to be easy targets for imposed change. He further quotes Robert McC. Adams: "The oversimplification of ideologies and their modes of attachment to other cultural phenomena may well be the distortion that most seriously threatens to undermine studies of ideologies in early states and civilizations" (Flannery 1999: 220). In this work we take up the challenge to resist the oversimplification of ideology, attempting rather to give it its proper due in the creation of the Merina polity of the eighteenth century.

### *The Gift of Speech, the Power of Words and the Monopoly of Discourse*

Many factors (including a large number identified by Flannery 1999) contributed to Andrianampoinimerina's political success: leadership skills, the help of talented advisors from Imerina and from the east coast of the island (political actors in their own right), military prowess and ruthlessness, shrewd and multiple marriage alliances, diplomacy, the circumvention of a convoluted system of class privilege, agricultural innovation, labor conscription, trade monopoly and market reform and control, etc. We want to draw the reader's attention, however, to one particular skill of Andrianampoinimerina, the gift of speech, mentioned in comments made by eye-witnesses of this historic figure and found in comments made by a secretary to the Merina court of the generation that followed Andrianampoinimerina. *Ombiasy*, ritual specialists, from the South-East were called upon as political consultants for numerous polities across the island (Raison-Jourde 1983: 32). Andrianampoinimerina in turn invited a delegation of Antemoro and Anakara to join the ranks of his local political consultants and ritual specialists. These Antemoro and Anakara advisors saw Andrianampoinimerina as not only good-looking, "bright faced," smart, but also as a gifted speaker (*mahay miresaka*)[2] (Kasanga 1956: 20). This reputation for oratorical skill on the part of Andrianampoinimerina was also noted by the indigenous historian, Raombana, born in 1809, one year before the death of Andrianampoinimerina. Raombana served as secretary to Queen Ranavalona I, daughter-in-law to Andrianampoinimerina, who succeeded to the throne after the death of her husband, Radama I. Raombana and his twin brother, Rahaniraka, were sent to England in 1820, by Radama I; this move took these children from their birthright world of the Malagasy language and orality and placed them into the literate world of the English. Intended to become linguistic mediators and culture brokers, Raombana and his brother were brought back to Madagascar in 1829 at the age of approximately 20 to serve Queen Ranavolona I. The brothers came back *oriented* in the fashion of the West, but *alienated* from their native language and cultural logic. Ayache's research on Raombana (1976) helps us understand Raombana as a disgruntled political servant who understood (and referred to) Andrianampoinimerina as a usurper (cf. Raombana 1980: 198–199, 202–203)

[2] *"Hitanay ny mpanjaka, olona vintanina, fa matanjabatana sy saina, mazava endrika, mahay miresaka. Olon-tsoa raha ny fijerinay azy, nefa toa saro-piaro sy tokam-po"* (Kasanga 1956: 20).

who had displaced from power the line of nobles from which Raombana was descended. However Raombana, in the journal he kept concerning Malagasy history, also recognized this "wily usurper" (1980, 203) as a "wily orator" (ibid: 178–179) according to the extant oral traditions. Additionally of interest is that we find embedded in Raombana's journal (1980: 35) the following remark concerning the population of Imerina:

> But what struck me most was their retentive memory, and the persuasive Eloquence which every Hova [Merina] people seem [sic] to possess from the old men to the mere children that plays [sic] and romp about the streets, and in the public assembly or Kabary, they display their Eloquence in the highest degree. They speak without pre-meditation, and have a great flow of words, and no confusions are at all perceived in them.

The *Tantara ny Andriana eto Madagascar* (Callet 1981) accords Andrianampoinimerina his share of clever, elegant, and moving discourse. Should we be taken in by this reputation for oratory on the part of Andrianampoinimerina in that Callet's informants were certainly partisans to power and privilege given their associations with and loyalties to this sovereign? Perhaps, but this reputation is additionally vouched for in the comments of the Antemoro and Anakara specialists that served at Andrianampoinimerina's court, specialists who were certainly well-versed themselves and well-traveled across the island. Are we personally enamored with this discourse? In some ways, yes, given the poetic voice and resonances of the discourse! There is an additional question, however, concerning whether this discourse can actually be attributed to Andrianampoinimerina either as skilful 'off-the-cuff' remarks of a 'wily' gifted orator, or as remarks prepared by him. Yet, even if such remarks were actually crafted by a ghost *mpikabary*[3] or *mpanolo-tsofina*[4] before the fact, this discourse is still striking in its cleverness and beauty. It is, however, the case that any and all of these scenarios point to the importance of royal discourse in the creation of the Merina state and during its early years of expansion beyond Imerina. It is important to understand the power of language in primarily oral societies. This is the "power to name," the power to interpret, and the power to delimit the terrain of

[3] A *mpikabary* is an individual who is called upon to perform public oral discourse for families in important matters such as marriages and funerals.

[4] In the Betsileo dialect this term, *mpanolo-tsofina,* can be translated as "who lends [some words] to the ear [of the one speaking]." In important discourse, a *mpikabary* is often flanked by one or two individuals who serve as aide-memoire, whispering into the ear of the *mpikabary* names and details that may have been missed or forgotten by the *mpikabary.*

discourse (Kus and Raharijaona 2002). This is a culture where many recite powerful verse such as *hain-teny*, a term that lends itself to multiple interpretations, including "science of language," "burning words," "life of words," and "breath of life" (Fox 1990: 40). Paulhan further helps us understand that *hain-teny* are not only recited, they are "made to fight" in public discourse. A century after Raombana's remarks, Paulhan (1938; 1991; 2007), in his studies of *hain-teny*, came to appreciate the power of oral verse in Madagascar to persuade, to dissuade, and to incite the imagination. This is a culture that has many additional named genres of speech acts including *isamanga* (blue/beautiful speech), *volamandiadia* (speech that tramples), and *fampariahitsa* (tossing of the seeds of the *ahitsa* plant; this last term carries the image of planting ideas for further reflective thought) (Michel-Andrianarahinjaka 1986). Further, there is the tradition of the *kabary* in the highlands, public discourse, which is in theory, and is made in practice, to resemble discursive exchange between a sovereign and his subjects. A contemporary Malagasy term for public, political discourse, *fampielezan-kevitra*, which westerners might gloss as "propaganda," carries as well the notion of using discourse to reveal and share ideas with the assumption that the audience is made up of sophisticated consumers of speech acts and capable of following [and assenting to] thoughtful analyses.

Only recently have political scientists taken the "linguistic turn" (cf. Lears 1985: 589; Norval 2000: 313–314) in their analyses of 'ideology', based on the argument that "ordinary language analysis" of contemporary political language reveals that the dominant ideology seeks to control and restrict the content of cultural concepts used in the language of political discourse. Echoing that professorial analytical turn, and most likely preceding that analytical turn, has come the profession of 'spin-doctoring'. We would argue that spin-doctoring had been a fully developed art form of the Malagasy *kabary* preceding and during the time of Andrianampoinimerina involving the struggle to control, not so much 'concepts' as, symbols, tropes and metaphors used in the construction of conceptual fields. Where oral discourse is a cultural art form practiced by the rank and file of the citizenry and, where nascent power is under construction, not yet capable of a monopoly of coercive force, then the art of persuasion is indeed a powerful and necessary tool of polity and hegemony building (à la Gramsci's (1985) appreciation of the concept). Consequently, we should not be surprised by the oft repeated assertion in the oral traditions of Andrianampoinimerina that the construction of public works (e.g., dikes, new rice fields) and the construction of the polity, itself, resulted from his words, the directives of his speech that were ultimately brought to realization by the labor of the people[5] (Callet 1981: 745).

There is a vast amount of official, public discourse that is attributed to Andrianampoinimerina in the *Tantara ny Andriana eto Madagascar* (Callet 1981). Not surprisingly that discourse covers much terrain in terms of topics and tropes. For the purpose of our discussion in this work, we want to bring the reader's attention to two powerful, recurring tropes in this discourse: (1) *ny tany sy ny fanjakana* (the land and the polity) and, (2) the necessity that it be a "singular" sovereign (*andriana tokana*) that reigns over *ny tany sy ny fanjakana*. These tropes proved central to the importunate crafting of political logic wherein the natural and the political order were intimately intertwined, and wherein the 'singularity' of the sovereign came to be confounded with the health and well-being of the citizens of the polity.

## Ny Tany sy ny Fanjakana: *An Indigenous Notion of the State*

During the time period when Andrianampoinimerina came to power (1787–1810), the polity was referred to as "*ny tany sy ny fanjakana*" (Callet 1981). This term has been classically translated as "the land and the kingdom" (*le pays et le royaume*). The first thing of note in this formulation is the seeming duality of "land" and "kingdom/rule." Such 'duality' in phrasing is neither uncommon nor awkward in the Malagasy language that is found in the *Tantara ny Andriana eto Madagascar* (ibid.). It is actually a linguistic ploy to signify complementariness and consequently singularity and completion, the creation of a larger whole. Other examples of duality signifying a larger whole are the terms *ny farany ny lohany* which can be translated as "[in] the end and [in] the beginning" but is meant rather to signify 'in perpetuity'; *vady aman-janaka* which can be translated as "spouses and children" but is meant to designate the 'family'; and *raiamandreny*, which can be translated as "father(s) and mother(s)" but is meant to designate 'elders.'[6] More poetic images of the creation of a

[5] "*Kanefa na andriana aza aho, hianareo vahoaka no nanana'ko izany'ny vava ko foana no manoro, f'hianareo mahavita; dia mankasitraka aho mankatelina, izay atoro ko dia vita nareo avokoa*" (Callet 1981: 745).

[6] See Kus and Raharijaona (2000) and Kus (2008) for an extended discussion of the "confounding" of the two capitals of Antananarivo and Ambohimanga by the singular sovereign, Andrianampoinimerina.

whole through duality can be found in *hain-teny*, one in particular is brought to mind concerning *vary sy rano*, "water and rice," never separated in the rice paddy and further conjoined when cooked in the rice pot (Paulhan 2007: 98), both images allude to the substance of life and the nature of love. Still today, to designate the territory of the Malagasy Republic, the appellation of "land and polity" is used both in political discourse and in official documents: *Ny Tany sy ny Fanjakana manerana ny Repoblikan'i Madagasikara* (The Land and the Polity across the entirety of the Malagasy Republic).

Let us look however at the two terms separately to better understand their relationship and their conjoining. One might begin by appreciating, in Peirce's fashion (1958–1960) '*ny tany*' or 'the land' as an iconic and indexical sign and '*fanjakana*' or 'polity' as a symbolic sign, an abstraction in need of fill. "*Ny tany*" in its first sense is "land, soil, the earth." Yet the dictionary of Abinal and Malzac will also reveal that this term can be used to designate "*pays, contrée, royaume*" (1955: 673). As among any "inhabitory people" (Snyder 1990), including skilled and knowledgeable agriculturalists, "the land, the earth" is a powerful symbol. More explicitly, more primarily, and perhaps even more correctly "the land, the earth" is a powerful iconic sign and a powerful indexical sign. Were space permitting, this section could be filled to the brim with the concrete and existential poetics of all that land and earth can mean to the Merina; as with naked feet they are grounded and as with naked hands they gather the fruits of their labor, and as they have and do move from the river valleys that nourish their rice crops, to the slopes where their cattle graze, to the heights that visually dominate their land, named Imerina, "Visible from Afar." As they bury the umbilical cords of newborns in their villages and as they bury their dead in tombs, the landscape serves as index of affiliation and status. Sensual engagement with the earth, with the land, also offers a wealth of concrete, poetic images for bricolage in the conceptual domains of local knowledge, domains that touch not only ontological and epistemological speculations, but intimate personal relationships as well.

Consider the following *hain-teny* (Paulhan 2007: 82):

| | |
|---|---|
| *Sira vatra avy any andrefana* | [I am] the coarse salt that comes from the West |
| *Tantely mandry avy any atsinanana aho* | I am the thick honey that comes from the East |
| *Andramo rankizivavy* | Taste [of me], young woman |
| *Sady fy no mangananganana* | Sweet, yet savory |

The earth and its offerings can portray a depth of sentiment and sensuality that, not surprisingly, shocked the early Protestant missionaries to the island. If the private can be so conceptualized with respect to the land, we will come to see below that so can the political.

Perhaps it is possible to speak of the term '*fanjakana*' as more immediately abstract and thus symbolic in Peirce's (1958–1960) sense of the term. The term "*fanjakana*" comes from the root "*zaka*" which can be translated as "that one can carry, bear the weight of, that one can govern, that which one is capable of" and also the meaning that "one can do with, or command according to one's will" (Abinal and Malzac 1955: 866). Certainly, this is an interesting range of meaning. So the term "*fanjakana*" as a noun can itself mean "administration" as well as "reign" with all the ambiguity of *order* and *exploitation*. The term "*fanjakana*" can be qualified. The term "*fanjakan'i Baroa*" or the "reign of the sovereign Baroa" is synonymous with political exploitation and oppression. The term "*fanjakan'i Bezanozano*" or the "administration of the Bezanozano (ethnic group)" is synonymous with anarchy. The term "*fanjakana*" is thus a field open for ideological filling. Perhaps we can pursue this filling along the lines of Flannery's assertion that a 'state' needs an ideology appropriate to it.

In the case of the Merina *fanjakana* we argue that its ideological filling involves an appropriation and reworking of the domain of *ny tany* with which it is co-joined. We need, however, to remember that we are dealing with sophisticated producers and consumers of systems of local knowledge that are philosophically rich and provocative, and to remember that we are dealing with sophisticated consumers of oral discourse that is linguistically subtle and symbolically dense. Hence, in early state-like polities (that lack coercive means to impose belief) state ideology is neither fortuitous nor gratuitous.

In the recorded oral traditions for the Merina (Callet 1981), the land, the earth, *ny tany* serves numerous times as icon and index of suitability to govern. In the beginning, the sons of the first chief of those who inhabited the land before the Merina set out looking for food. They came across a hedgehog. The younger son offered the choice to his elder brother: the land or the hedgehog. The elder brother was content to satisfy his immediate, individual needs (Callet 1981: 12). The right to rule went to the junior. Foucault is said to have argued that all power needs resistance for its existence and its exercise (Dreyfus and Rabinow 1982: 147). Andriamasinavalona was the great great grandfather of Andrianampoinimerina. He was also the sovereign who is credited with bringing an original unity to the Merina polity, a unity that had been envisioned by his predecessor, Ralambo (Callet 1981: 147) who conferred the name, Imerina, on the nascent polity. Andriamasinavalona and his advisor Andriamampandry seemed to have shared Foucault's understanding of power. When Andriamasinavalona was contemplating the large public work of draining and diking the vast marshland of the Betsimitatatra in order to create new rice fields to nourish the population, he was concerned that the population would refuse to follow his directives. He and Andriamampandry conceived of a ploy to gain the people's consent in this major undertaking. Andriamampandry told Andriamasinavalona to bring a dry, breakable stick to Andohalo where Andriamasinavalona was to assemble the people to receive his public address/directives or *kabary*. (Andohalo would serve as location for royal *kabary* through the days of the Merina dynasty, colonial discourse during French occupation, and even Presidential address into the time of Tsiranana (1958–1972) the first president of Madagascar.) As agreed upon, Andriamampandry planted himself in the audience. When Andriamasinavalona unveiled his plan, Andriamampandry rose in protest to say: "No! The people refuse to undertake this corvée labor." Andriamasinavalona, as agreed upon with Andriamampandry, rose in feigned anger and chased and beat Andriamampandry until the people cried out to spare the counselor claiming that he was not speaking on the people's behalf. The men of Imerina would indeed undertake the great project of constructing the dike named *Vahilava* (Long [and straight] dike) so that Andriamasinavalona would be able to nourish their wives and children (*hanao izay hahavoky ny vadi-aman-janak'ay*). What is most interesting is that this story appears as a footnote (the equivalent perhaps of an editorial/informant's clarification) in the lengthy section of the oral traditions (Callet 1981: 297) describing Andrianampoinimerina's construction of dikes, including the re-enforcement and reconstruction of the *Vahilava*, Andriamasinavalona's original project. It is in this section describing Andrianampoinimerina's exhortations to the population (*kabary*) that we find once again an instance where he claims that his *words*, the directives of his speech, were ultimately brought to realization by the labor of the people. One might say that we have an interesting contrast here between Andriamasinavalona "carrying a small dry breakable stick" and Andrianampoinimerina speaking loudly, but in poetic tones, of ultimate respect and concern for his charges. It is also in this section that Andrianampoinimerina claims that his only enemies are *rano vaky*, or the water that breaks (the dikes), *afo* or fire, *mosary* or famine, *havandra* or

Figure 2. The flooded Betsimitatatra on February 19, 2007 as seen from just below the Rova of Antananarivo.

hail and *rivotra* or wind. This leads us back to Foucault's argument that all power needs resistance. Does it matter what form that resistance takes and how the enemy is conceived?

In order to understand the socio-political rhetoric of the official royal oral traditions (Callet 1981) concerning Andrianampoinimerina's rise to power it is necessary to look at the earlier historical scenario that set in place the pragmatic and ideological props for this move. Some four generations or so previous to Andrianampoinimerina's rise to power the Merina state was unified by the then reigning sovereign, Andriamasinavalona. As the historical tale runs, Andriamasinavalona could not and would not choose a successor among his four sons, despite the fact that there were traditional means to do so. For instance, one could observe how the potential heirs arranged themselves in a room and then choose the one who had the habit of leaning against the central pillar (ibid.: 406). Or one could make the potential heirs sleep in the same room and, in the middle of the night enter the room with a light and see who slept with his head raised (ibid.: 302). Not only would Andriamasinavalona *not* follow the indications of these tests to choose a single heir, he also refused to heed the advice of his counselor, Andriamampandry. (It is not without significance that this counselor's name translates as the "Noble-that-creates-calm/peace.") This counselor warned Andriamasinavalona several times, and in very dramatic fashion, of the consequences of such a refusal. For instance, Andriamampandry showed up at the royal residence one day carrying a basket that contained four birds of prey (*voromahery*). When released these birds flew around and stirred up enough soot and dust to cover Andriamasinavalona and fill his eyes and throat and cover dishes of rice making his meal inedible. The 'Peace-maker' commented that the incident was only a metaphor for what would happen in Imerina if it were divided into four parts (Callet 1981: 300). Even more dramatic is the time that Andriamasinavalona assembled his people as advised by the 'Peace-maker'. The 'Peace-maker' then let loose a bull among the crowd. Many were trampled and injured, and some pregnant women aborted. Accepting responsibility before the outraged sovereign, the advisor explained that the destruction and loss of life that resulted from his action was negligible compared to what would happen to the Merina if Andriamasinavalona let four individuals exercise royal authority (ibid.: 302). Unfortunately, Andriamasinavalona stubbornly refused to designate a single heir, and so it came to pass that war and famine continuously plagued the divided state (ibid.: 300, 310–13).

Several generations later Andrianampoinimerina laid claim to sovereignty over the four districts of the divided state. The bases of his claim as presented in the recorded tradition (Callet 1981) are too extensive and involved to present here. Suffice it to say that these bases included not only a correct genealogy, but also the fact that Andrianampoinimerina liked to lean on central pillars and to sleep with his head raised. Perhaps, given the concerns of this chapter (and this volume), the most significant epiphany of Andrianampoinimerina's predestination as singular heir to Andriamasinavalona's united polity is to be found in a tale of Andrianampoinimerina's maternal grandfather, Andriambelomasina. This grandfather was confronted by that recurrent problem of all sovereigns: who to designate as heir. According to oral tradition, Andrianampoinimerina was summoned along with other possible heirs to the polity of Andriambelomasina. There laid out before all the contenders was a clean mat covered with silver, jewelry, semi-precious stones and other sumptuary items. All summoned were allowed to choose. Andrianampoinimerina, with determination and without hesitation, took the basketful of dirt that also lay on the mat (choosing the land rather than the equivalents of the hedgehog of his earlier ancestor). His grandfather Andriambelomasina consequently declared: "*lasa'ny ny tany sy ny fanjakana*" literally "[he] took the land and the polity", but figuratively indicating that Andrianampoinimerina had been designated by fate to detain authority and power (Callet 1981, 405).

While genealogy and omens seem to have been clear in designating Andrianampoinimerina as heir to the state created by Andriamasinavalona, the exercise of this claim came through diplomatic acts and physical battles. These acts, according to the oral traditions (Callet 1981) were all directed toward the explicit goal of reuniting the divided state. It is the symbol of unity that plays a central and critical role in the ideology/logic that Andrianampoinimerina used to legitimate his power and politics as described in the oral traditions.

### *The Land Bears the Imprint of the Polity*

The land bore physical witness of the '*fanjakana*'s' existence. Collective diking and draining projects brought the land into increased productivity. Indeed, under Andrianampoinimerina, the draining and diking of the vast plain of the Betsimitatatra that partially surrounds Antananarivo allowed for a rich second cropping of rice, a counter seasonal crop attesting to the audacious power of the state. This *second* crop of rice became the most appropriate *first* fruit offering (*santatra*) of rice to the polity under Andrianampoinimerina's reign (Callet 1981: 60–61). The same ditches that turned the Merina plains into a rich rice basket, served also as political boundaries. Further, previous to Andrianampoinimerina's rule, internecine warfare often involved the destruction of these ditches causing horrific famines. The names that were given to these famines such as, "Sunken eyes" and "To assuage children's' hunger with promises" bear testament to their horror. (See Table 1.) Thus harvest and famine become indexes of governance. When Andrianampoinimerina said that "my only enemy is famine" (*fa ny mosary no fahavaloko*) (cf. Callet 1981: 803) and "my only equal is rice" (*ka izaho manao ohatra hoe izaho sy ny vary no mitovy*) (cf. Callet 1981: 746), and subsequently claimed that the role of the *fanjakana* and its ruler is to render the land productive, his words did not necessarily ring hollow. They were also words that could be 'indexed' against experience.

The territory of Imerina not only 'bore the fruits' of governance, it also bore the conceptual overlay of the *fanjakana*. Today one can still look out from the height of Ambatomiantendro, whose name can be interpreted as

Table 1. The seven named periods of famine preceding the reign of Andrianampoinimerina.

| NAME OF FAMINE | TRANSLATION |
|---|---|
| Imavovava | Yellow Mouth (a physical testament to hunger) |
| Imiambohohobe | (When) Many Turned their Backs (or Died) |
| Imifontsamilahy | To Devour a Banana (with no regard for etiquette) |
| Imamahantsofinjaza | Feeding the Ears of the Children (with only promises of nourishment) |
| Elatrakoho | (To lift the) Wing of the Chicken (to see whether the yellow color of the skin reveals that it is fat) |
| Itsimiofy/Ihomantsimivaofy | (One) Does not wait to Peel (before eating) |
| Ilalimaso/Irangasofina | Sunken Eyes/Ears that are Detached from the Jaw (images of extreme malnutrition) |

both "The boulder that surpasses all others [in height]" and "The height where all peaks meet," to view the dikes and ditches that have been created by *fanjakana* of the past centuries. It was at Ambatomiantendro that Andrianampoinimerina publicly revealed his intentions to unite the country of Imerina. Andrianampoinimerina 'grounded' the ideology of the Merina polity in the local topography and in the fruits that the land bore. This was an ideology crafted around relatively abstract notions of singularity, completeness, perfection and the endurance of the sovereign and the polity. Upon unification and pacification of Imerina, Andrianampoinimerina announced his expansionist intentions by claiming that the sea would be the limits of his rice field. When he chose his successor he did so by handing him a basketful of dirt, *ny tany*. Callet spoke with an elder who had seen the last days of the reign of Andrianampoinimerina. This individual, he wrote, held up a small pinch of dirt and remarked about Andrianampoinimerina that "no clump of dirt [as small as the one I am holding] was untouched by his thoughts or his laws" (Callet 1981:746).

*Ny tany sy ny fanjakana* is still used in current political discourse. To signal their support when Ravalomanana claimed presidential authority, rural agriculturalists during one political address/*kabary* brought to the then new president, traditional offerings of *santatra* which included rice, milk and honey to symbolize 'abundance,' guinea fowl as a sign of 'loyalty,' quail as a sign of 'stability,' eel as a mark of 'vigilance,' and crayfish as a sign of 'firmly established power' (cf. Kus and Raharijaona 2006a: 246).

*Ny tany sy ny fanjakana*, written as five words and two different nouns, is an artifact of French and English missionary transcription of the Malagasy language into Roman script. In practice, in this agglutinative language, it is spoken as one word. The intimate connection of 'land' and 'governance' not only underpinned the logic of Andrianampoinimerina's claim to "have no equal except rice" (Callet 1981: 746) but it brought poetic force to the claim as well. While having no rival, Andrianampoinimerina did claim to have enemies, of which famine was only one.

### *Power and Resistance that Allows for Existence*

In direct discourse as presented in the *Tantara ny Andriana*, Andrianampoinimerina argued quite explicitly that political unity is the only way to bring peace and prosperity to a state plagued by disorder, famine and warfare. The way to guarantee such peace is to have a single head of state (Callet 1981: 526). Andrianampoinimerina's arguments for unification under a single sovereign, namely himself, were reinforced copiously in symbolic political acts and metaphors made material (cf. Kus and Raharijaona 2006a). We will focus on only a few examples where the symbolic significance is quite obvious; in particular, we will focus on the litotes that contrasts the number "two" with a concept of 'singularity'. One of the royal residences at the capital of Antananarivo is called Masoandrotsiroa, the "Sun-that-is-not-two". In similar fashion one royal fetish or palladium is called Manjakatsiroa or the "Ruler-that-is-not-two". At times various citizens were put through a trial-by-ordeal (*tangena*) to prove their loyalty to the state. They were required to swallow a potentially poisonous substance while repeating an oath: "If we attempt to create/promote a second sovereign let us perish,"[7] (Callet 1981: 1044). Perhaps the symbolic quality of the number "two" is seen even more clearly in the statement of a prime minister when he pledges his loyalty to

---

[7] *Raha hanao andrian-droa amin'ny tany dia matesa.*

the sovereign. He says that if he is guilty of treason then he should be cut into two equal pieces (*roa mitovy*) and thrown to the dogs (Callet 1981: 393). The argument here suggests that the idea of *duality*, when placed in relief by the notion of *singularity* in the Malagasy context, carries the connotation of competition rather than opposition; competition offered by a rival, not opposition offered by an enemy. At times one recognizes such rivals of Andrianampoinimerina through symbolic gestures that usurp royal privilege. For instance one rival, Andriamary, approached Andrianampoinimerina carrying a red parasol, a privilege reserved for the sovereign (ibid.: 549). Such rivals are thus offering themselves as alternative candidates for the same political position. They are not offering alternative political platforms or philosophies (that is why they come carrying red parasols rather than little red books of the revolution.) Yet Andrianampoinimerina does declare several times in the *Tantara ny Andriana* that he does indeed have enemies, *fahavalo*, the most dangerous of whom is famine. "I have no enemies except famine"[8] (ibid.: 803). If we pursue the cultural logic that serves as the context of the socio-political ideal of unification we can understand how the claim by Andrianampoinimerina to have only "natural disasters" such as flood, famine, fire, hail, and wind (cf. Callet 1981: 745, 801) as enemies is not contradictory with the fact that he engages in numerous armed confrontations with nobles and social groups in his attempts to bring them under submission.

Talking in the language of political rhetoric, Andrianampoinimerina is often quoted justifying his actions by claiming that they are means to political unification. Why unification? This is because peace and prosperity are attendant upon unification; without unification, chaos and insecurity reign. Fear of brigands and continuous fighting between petty 'kingdoms' keep people from working in the fields and rice paddies and keep them away from markets. Political insecurity induces famine which in turn encourages "the big to take advantage of/'eat' the little and the little to steal" (*ny lehibe mitady hihinana ny kely ary ny majinika mitady hangalatra*) (Callet 1981: 802). This is, in fact, the claim that is made for the period when Andriamasinavalona left a divided kingdom. One famine is said to have lasted seven years.

Andrianampoinimerina was always claiming to wage battles, as well as direct social labor in projects such as diking and draining marshland areas, not for any personal gain but for the welfare of the populace. In one piece of discourse he lays out this logic (Callet 1981: 809):

---

8 "...*f'izaho tsy mba manana fahavalo fa ny mosary no fahavalo ko.*"

> ...the core of the state ultimately is based on the work that renders the land productive since this is the way to insure the well-being of the populace. I am happy to see your bright faces and foreheads without wrinkles (*madio-dio tava sy tsy miketron-kandrina*)...because you are well-nourished, I am happy to see that you are healthy and strong: my desire has found its realization.

From this ideology/logic of a social contract wherein political power is only legitimately held and exercised for the benefit of the population, it follows that Andrianampoinimerina can make the claim that his only real enemies/rivals are *rano vaky* or the water that breaks (the dikes), *afo* or fire, *mosary* or famine, *havandra* or hail and *rivotra* or wind. Thus people are not the ultimate enemy. They can become accomplices of these *real* enemies when they perpetuate political chaos (by supporting political rivals) or when they refuse to act in united social effort to support the welfare of the state. This logic of a socio-political order as ultimately interfaced with, if not inseparable from, a natural or cosmological order is further elaborated on in Andrianampoinimerina's social policy regarding the confrontation of these public enemies. These public enemies are to be fought through united *public* effort. Consequently there was to be public upkeep of dikes. In case of fire all citizens were to answer cries for help. Industriousness on the part of all citizens, nobles as well as commoners, would insure abundant harvest. For hail and wind, which are a bit trickier to handle, Andrianampoinimerina also had a *social* solution. All citizens would pay a tax to support full-time specialists to manufacture amulets against such destructive natural forces (Callet 1981: 801–3).

The logic that confounds natural and social orders, that confounds natural and political enemies, finds its complement in Andrianampoinimerina's remark that he has only one friend, one equal, *rice*, the staple crop of the Malagasy highlands (ibid.: 764). Rice and Andrianampoinimerina united together (as are *ny tany sy ny fanjakana*) become the basis of the existence and the prosperity of the Merina people.

The series of metaphorical equations laid out above allowed for the crafting of a further powerful trope to be used for political ends. Andrianampoinimerina claimed that the sea was the border of his rice paddy[9] (cf. Callet 1981: 433, 493, 803). This claim was continued by his son, Radama I, and it was brought to partial fruition by this son and his

---

[9] See Kus and Raharijaona (2006b: 318–20) for a fuller and nuanced discussion of the two alternative phrasing of this claim: (1) *Ny ranomasina no valam-parihiko*, (2) *Ny riaka no valam-parihiko*.

successors, who eventually brought one-third of the island under Merina political control.

## *Conclusion*

> *Itataram-poza; iasana tanety, avoko; anarahana andriana, hena.* (To dig into the small channels [in the rice field after the harvest] one will find small freshwater crabs [to eat]; to dig into the soil one will find the roots of the *avoko* [to eat]; to follow the sovereign will bring one meat [to eat]) (Houlder 1960: 118).

In the political discourse attributed to Andrianampoinimerina, one is not dealing with the cultural concept of an enemy that is most familiar to the modern Western world, a concept of an opposite, an arch-enemy. Therefore it is not surprising to note that Andrianampoinimerina often asked his *enemies* to submit and accept him as their father. Nor is it surprising that the terms *fahavalo* (enemy) and *rafy* (rival) are not used in the *Tantara ny Andriana* (Callet 1981) to designate reciprocal relationships. That is, while Andrianampoinimerina has enemies and rivals, he is not referred to as the rival or the enemy of others. Of course this is political rhetoric! We are not claiming that Andrianampoinimerina exercised his power benevolently nor that there were not bodies strewn on battlefields and families sold into slavery. What we are suggesting is that in the comparison of Merina ideology (and even some, but not necessarily the most recent, contemporary Malagasy political rhetoric) to modern Western forms there is a question that poses itself. What are the pragmatic consequences, if any, of these two alternative logics of political resistance and the constitution and legitimation of power structures? That is, do these two alternative conceptualizations place limits or extend limits on the form and exercise of political power? Or perhaps more importantly, how do they affect the possibility of resisting and ultimately dismantling power structures that form the bases of state organizations?

Certainly, one form of resistance is powerful poetic political commentary in multiple forms (e.g., *ohabolana*/proverbs, *hain-teny*, songs) used by generations of Malagasy, forms of the continuing "persuasive Eloquence which every ... people seem to possess from the old men to the mere children that plays and romp about the streets" (Raombana 1980: 35). During a series of field seasons in the 1980's, we witnessed firsthand in the highlands of the island the terrible suffering, both physical and mental, that political insecurity and exploitation can visit upon rural populations

(Kus 1997: 207–209). Yet, we also recorded many Betsileo *rija* (a form of sarcastic commentary on the quotidian put into song) from a number of different villages that, with pointed and poignant humor, spoke to the relationship of the land and the state.

Consider the following verse recorded at Mitongoa, Betsileo in July 1984.

| | |
|---|---|
| Metimety ny manam-basy | It is very useful to have a gun |
| Manam-basy, manan-kery | To have a gun is to have might |
| Manan-kery, manan-tany | To have might is to possess land |
| Manan-tany, mana-mpanjakana | To have land is to possess the state |
| Fa manan-tsoa ny manan-kavana | But to have family is to possess "the good" |

This traditional Betsileo *rija* was later arranged and recorded on CD by the band of Rakoto Frah, Feo-Gasy, in 1995, testifying to the continued relevancy of such verses to the Malagasy political context.

It was said of Andrianampoinimerina that his claim to authority over Imerina was presaged by the act of choosing a basket of dirt among the surrounding dazzling riches laid before him by his uncle: *lasa'ny ny tany sy ny fanjakana* ("to him belongs the land and the authority"). Consider the powerful words of the following Betsileo *rija* tape recorded at Andrainjato, Betsileo in May 1987.

| | |
|---|---|
| Any Andondona | [The sacred heights of] Andondona |
| Ny sape novolen'ny Foré | Have been reforested in pine trees [by the Ministry of Water et Forests] |
| Any Antalatan'Ampano | [The cascade of] Antalantan'Ampano |
| Ny rano lasan'ny Jirama | [Its] waters are possessed by Jirama [state owned Utility of Water and Electricity] |
| Any Andrainjato | At Andrianjato |
| Ny tany lasa "very tsy hita" [université] | The land is "lost to sight" [a subtle play on words: "lost to the University (of Fianarantsoa)"] |

In a poetic coup, in a final verse of this *rija*, using an implied/understood play on words, *lasany ny fanjakana ny tany* (the land has been seized by the state) our attention is brought to how land has been, and unfortunately continues to be, co-opted by the modern *Ny Tany sy ny Fanjakana*.

## *Bibliography*

Abinal, A. and V. Malzac. 1955 [1888]. *Dictionnaire Malgache-Français*. Paris: Editions Maritimes & Coloniales.

Ayache, S. 1976. *Raombana l'historien (1809–1855): Introduction à l'Edition Critique de son Œuvre*. Fianarantsoa, Madagascar: Librarie Ambozontany.

Callet, F. 1981 [1873–1875]. *Tantara ny Andriana eto Madagascar, Tome I and II*. Antananarivo: Imprimerie Nationale.

CNN News online. January 13, 2010. "Pat Robertson says Haiti paying for 'pact to the devil." http://www.cnn.com/2010/US/01/13/haiti.pat.robertson/index.html (accessed 14/10/2010)

Delivré, A. 1974. *L'Histoire des Rois d'Imerina: Interprétation d'une tradition orale*. Paris : Klincksieck.

Dreyfus, H. and P. Rabinow. 1982. *Michel Foucault, Beyond Structuralism and Hermeneutics*. Chicago: University of Chicago Press.

Flannery, K. 1999. "Process and Agency in Early State Formation." *Cambridge Archaeological Journal* 9(1) 3–21.

Fox, L. 1990. *Hainteny: The Traditional Poetry of Madagascar*. Cranbury, New Jersey: Associated University Presses.

Gramsci, A. 1985. *Selections from Cultural Writings*, eds. David Forgacs and Geoffrey Nowell-Smith. Transl. William Boelhower. Cambridge: Harvard University.

Houlder, J.A. 1960. *Ohabolana ou Proverbes Malgaches*, transl. M.H. Noyer. Antananarivo, Madagascar: Imprimerie Luthérienne.

Kasanga, F. 1956. *Tantaran'ny Antemoro Anakara teto Imerina tamin'ny andron'-Andrianampoinimerina sy Ilaidama*. Antananarivo: Société d'Imprimerie Antananarivo.

Kus, S. 1997. "Archaeologist as Anthropologist: Much Ado about Something after All?" In New Approaches to Combining the Archaeological and Historical Records. *Journal of Archaeological Method and Theory*, 4(3/4): 199–213.

Kus, S. 2008. "Appendix E Ambohimanga: Creating a Capital and a Polity in Thought, Deed and Dirt," in *Early State Formation in Central Madagascar: An Archaeological Survey in Avaradrano*, ed. Henry T. Wright. Ann Arbor: Museum of Anthropology University of Michigan Memoirs. 43: 301–311.

Kus, S. and V. Raharijaona. 2000. "House to palace, village to state: Scaling up architecture and ideology." *American Anthropologist*. 102(1): 98–113.

Kus, S. and V. Raharijaona. 2002. "The power to name and the claim to dominion of a Malagasy sovereign," in *The Dynamics of Power*, ed. Marie O'Donovan. Carbondale, Illinois: Center for Archaeological Investigation, Southern Illinois University, 362–83.

Kus, S. and V. Raharijaona. 2006a. "Materials and Metaphors of Sovereignty in Central Madagascar," in *Rethinking Materiality: The Engagement of Mind with the Material World*, eds. Elizabeth DeMarrais, Chris Gosden and Colin Renfrew, 239–48. Cambridge: McDonald Institute for Archeological Research.

Kus, S. and V. Raharijaona. 2006b. "Visible and Vocal: Sovereigns of the early Merina (Madagascar) State," in *Archaeology of Performance: Theaters of Power, Community, and Politics*, eds. Takeshi Inomata and Lawrence S. Coben, 303–329. Lanham, Maryland: Altamira.

Kus, S. and V. Raharijaona. 2008. "Desires of the heart" and laws of the market place: Money and poetics past and present in the highlands of Madagascar, in *Dimensions of Ritual Economy, Research in Economic Anthropology*, eds. E. Christian Wells and Patricia McAnany, Vol. 27: 149–185.

Lears, T.J.J. 1985. "The Concept of Cultural Hegemony: Problems and Possibilities." *American Historical Review*. 90(3), 567–593.

Michel, R. January 18, 2010. "*Lettre ouverte à M. le « Pazenkor-Président*." http://www.madagascar-tribune.com/Lettre-ouverte-a-M-le-Pazenkor,13389.htm (accessed 18/01/2010).

Michel-Andrianarahinjaka, L.X. 1986. *Le système littéraire betsileo*. Fianarantsoa, Madagascar: Ambozontany.
Norval, A. 2000. "The Things We Do with Words – Contemporary Approaches to the Analysis of Ideology." *British Journal of Political Science*. 30(2): 313–346.
Paulhan, J. 1938. *Les Hain-teny*. Paris: Gallimard.
Paulhan, J. 1991. *Hain-teny Merina: Poésies Populaires Malgaches,* transl. Jean Paulhan. Antananarivo : Société Malgache d'Edition.
Paulhan, J. 2007. *Les Hain-Teny Merinas, Poésies Populaires Malgaches*, transl. Jean Paulhan. Paris: Geuthner.
Peirce, C.S. 1958–1960. *Collected papers*. Vols. 1–8, eds. Charles Hartshorne, Paul Weiss, and A.W. Burks. Cambridge: Harvard University Press.
Raison-Jourde, F. 1983. Introduction to *Les Souverains de Madagascar: L'histoire royale et ses résurgences contemporaines*, ed. Françoise Raison-Jourde, 7–68. Paris: Karthala.
Raombana. 1980. *Histoires 1*, ed. and transl. by Simon Ayache. Fianarantsoa, Madagascar: Librarie Ambozontany.
Rushlimbaugh. 2010. "Obama Leaps into Action on Haiti." http://www.rushlimbaugh.com/home/daily/site_011310/content/01125106.guest.html accessed January 14, 2010.
Snyder, G. 1990. *Practice of the Wild*. San Francisco: North Point.

# FOREST DEPLETION IN IMPERIAL MADAGASCAR, c.1790–1861

Gwyn Campbell[1]

It was traditionally believed that most of Madagascar was originally covered by forest which, in the central highlands, humans had destroyed (see e.g. Aubert et al 2003). This article argues that, contrary to the conventional view, the central highlands possessed considerable woodland until the eighteenth century, and that both highland and eastern woodlands were rapidly depleted in the period 1790–1861 due chiefly to the Merina crown and its adoption of autarkic policies.

Early Western visitors were impressed by Madagascar's eastern forest: In 1820, British Agent, James Hastie, and LMS missionary, David Jones, noted "the immense forest of Alamazaotra" (Ellis 1838: 222) which missionary David Griffiths (who arrived in 1821) stated was "over forty miles wide... hundreds of miles in length... [covering] approximately one third of the land surface" (Griffiths 1843: 12).[2] In fact, the Tamatave-Antananarivo route crossed two forests: the Alamazaotra and Analamaizina ("Dark Forest"), west of which visitors were as struck by the denuded plateau. For example, on reaching Imerina's eastern frontier (Ankaramadinika), William Ellis (1862) observed that the country had few trees and within an hour's march (Anosiarivo) was "entirely destitute of wood" (Ellis 1867: 24; see also ibid, 22–3).

Recent research indicates that while much of highland and western Madagascar possessed a mosaic of natural vegetation, including ancient stretches of savannah, and human occupation appears to have led to firing of woodland notably from c.1100–1300 (See e.g. Bond et. al. 2008; Wright 2007: 10), scholars have emphasized that Andrianampoinimerina (r. c.1787–1810) introduced what one author calls "the first known attempt at woodland conservation anywhere in the pre-colonial African realm"

---

[1] This research was supported by the Social Sciences and Humanities Research Council of Canada.

[2] Joseph Freeman and William Ellis use much the same description, probably based on Griffiths' original manuscripts—see Freeman 1830; Ellis 1838. vol.2: 35. Others give different names for the forest, such as "Fanghourou" instead of "Analamazaotra"—see e.g. Descartes 1846: 250.

Figure 1. Travel through the *Alamazaotra* or Great Eastern Forest.
Source: William Ellis, *Three Visits to Madagascar* (John Murray: London, 1859), 319.

(Gade 1996: 106). Andrianampoinimerina prohibited exploitation of woodland on the twelve sacred hills of Imerina (Analamanga, Ambohidrabiby, Alasora, Imerimanjaka, Antongona, Antsahadinta, Ambohimanga, Ilafy, Namehana-Ambohidratrimo, Amboantany, Ambohijoky and Ikaloy), proclaiming: "the forest is a common national heritage from which orphans, children, women and others may live, which provides that which is necessary to the construction of our houses... where the clothes of orphans and the poor are [to be] found" (Callet 1974: 612). It is commonly held that Andrianampoinimerina's successors maintained this ban, in part because the forest formed a barrier to foreign invasion (Radama is reputed to have said that "generals" "Hazo and Tazo"—"Forest and Malaria"—were a match for any European army; Baron 1890: 198).

### *The Impact of Cultivation*

There is evidence that, prior to the late eighteenth century, Imerina possessed considerable hilltop woodland. The site on which Antananarivo stands was originally called "Alamanga" (*ala* "forest"; *manga* "blue") (Tacchi 1892: 484) and many other place-names, such as Analabe, Andohala, Analatsara, and Analanomby, similarly denoted the former presence of woodland (Gade 1996: 104). In Imerina, wet-rice cultivation dominated lowlands until the late eighteen century when, due to famine and population pressure, lower concave hillsides were cleared of growth and brought under cultivation (Mayeur 1777: 162–3, 171–2; Grandidier 1908: 79–80 and 1928: 3, 5, 9, 30–31, 40; *Robequain* 1958: 275; Raombana 1853b: 24, 26–7; Oliver 1886. vol.1: 250–1 and vol.2: 53; Decary 1947–48: 36–7). By the 1780s, only enclaves such as the Ankaratra Mountains possessed extensive woodland (Mayeur 1777: 165, 170). Andrianampoinimerina prohibited tree cutting on Imerina's twelve sacred hills but elsewhere, in order to promote agriculture, encouraged deforestation and forbade tree planting (Callet 1974: 730–1). Under Radama I (r. 1810–28) forest-firing to create land for the cultivation of haricot beans, maize, tobacco and rice was widespread. And whereas Ranavalona I (r. 1828–61) proclaimed "If the people cut and burn my forests, their wives and children will be reduced to slavery, for forests that are cut down or burned are destroyed" (Callet 1974: 612, fn.8), the evidence indicates that she too permitted large-scale forest exploitation (Raombana 1853b: 18–19).

Human intervention indubitably caused depletion of the eastern rainforest which supplied trees for Betsimisaraka canoes, and ships for the French Mascarenes;[3] and land for *tavy* (peasant swidden agriculture) under which forest plots on level or gently sloping land were fired and planted with rice for five to six years until diminishing yields led to

---

[3] e.g., Labourdonnais (1746) and D'Aché (1759) constructed fleets to attack the British from trees growing in Antongil Bay – Anon [c.1790]: 60, 85–7.

relocation and clearance at a higher level (Couland 1973a: 155, 159–60, 161–7, 169, 172; Hugon 1808: 9; Chapelier 1940: 51; Michel 1957: 166–7; Le Sage 1816: 36; Grandidier 1928: 13–14, 16, 51).

In addition, some permanent settlements formed within the forest: Nicolas Mayeur (1785) found rice, sweet potato, sugar cane and banana cultivated at Ampasimpotsy; and Hastie sizeable rice fields at Marozevo (Mayeur 1785: 203; Hastie 1817). However, such permanent clearances were limited. Indeed, Dumaine found not only tavy-induced destruction but also deserted agricultural settlements (Dumaine. 1790: 254–5, 256–9). From 1817–30, first Betsimisaraka-Betanimena warfare and subsequently Merina military expansion caused high mortality and population dispersion amongst peoples of the eastern forest (Hastie 1817: 189–189b, 191–3; Hastie 1822; Raombana 1853a: 235–6; Raombana 1853b: 122, 137). Again, in the late 1830s, the forest village of Marozevo was abandoned when Ranavalona subjected its inhabitants to the *tangena* (poison ordeal) (Campbell 1963: 471). The forest became home to bands of refugees from formerly settled communities, including plateau peoples fleeing military

Figure 2. *Tavy*, Ifasina.
Photo reproduced by permission of Rhett A. Butler, ifasina0172 – http://travel.mongabay.com/madagascar/images/ifasina0172.html (19/05/10).

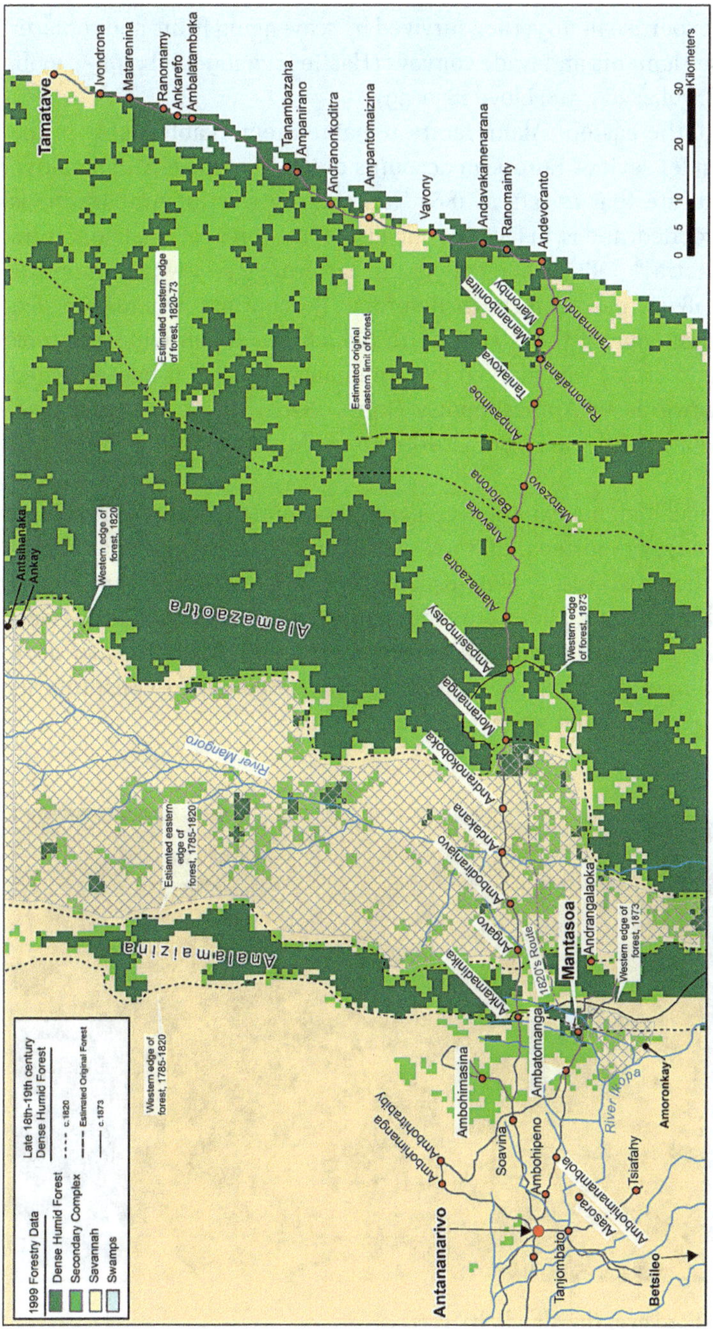


Map 1. Antananarivo-Tamatave route, late 18th to late 19th century.
Source: Gwyn Campbell, *David Griffiths and the Missionary "History of Madagascar"* (Leiden: Brill, 2012), 617.

and civilian *fanompoana* (unremunerated forced labour for the state) but few had expertise in *tavy*; they survived by scavenging fruits and roots, or raiding settlements and trade convoys (Hastie 1817: 189–189b, 191–3, 196b; Ellis 1838. vol.2: 369, 518; Lloyd 1850: 59).

Indeed, the eastern Alamazaotra remained remarkably stable in the period under review. European accounts of the Tamatave-Antananarivo route indicate that as late as 1862 "the country west of Ampasimbe is thickly wooded, and is, in fact, the outskirts of the forest of Analamazaotra [Alamazaotra]" (Oliver 1881: 154–5), the eastern edge of which, from the 1820s to 1870s, passed through Beforona. By contrast, whereas in 1821 the western edge of the Alamazaotra (which was 64km wide and took two days to cross) was marked by Moramanga, by 1873 it had retreated 14km eastwards to Ampasimpotsy (Jones 1820a; Hastie 1821 and 1822; Johns 1826; Griffiths 1843: 89; Lloyd 1850: 71; Mullens 1875: 37; Campbell 1963: 471).

The Analamaizina experienced earlier and more extensive shrinkage. Mayeur in 1777 reported that it was about 100 km from Antananarivo and took two hours to cross (7.8km–8.5km). However, only eight years later he reported it to be a 1.5 hour crossing (5.85km–6.4km) between the rivers Mandraka and Ankamadinika (Mayeur 1777: 186; Mayeur 1785: 207). Andrianampoinimerina considered the River Betatao to constitute its western border (Callet 1974: 667).

Figure 3. Ford over the Mandraka, 1873.
Source: Joseph Mullens, *Twelve Months in Madagascar* (London: Nisbet, 1875), 40.

Other traditional activities that impinged on forest resources were the production of dyes to colour cloth, mats and baskets, and salt. Dyes were produced from a large range of vegetable and mineral sources, including red dye from the bark of the *nato*; and blue dye from the bark of the *somangy* / *somontsohiny* (*Kigelianthe madagascariensis* Spreigne) and the pulpy aril surrounding the seeds of the Traveller's Palm (*Ravanala madagascarensis*) (Campbell 2005: 33; Baron 1890: 208). Again, Imerina imported prime salt from the west coast and less esteemed black salt from the ashes of reed mace (*Typha communis*) found in the eastern forest and Ankay basin (Mayeur 1777: 176 and 1785: 227; Dumaine 1790: 276; Chapelier 1940: 81; Grandidier 1908: 361, 366 and 1928: 199–200, 214–5, 295; Chapus and Birkeli 1944–45: 69; Hilsenberg and Bojer 1833: 253). However, the use of forest resources to produce dyes and salt was too restricted to have caused significant deforestation.

Indeed, it is here argued that, rather than traditional activities such as *tavy* and salt extraction, polices associated with the rise of imperial Imerina were chiefly responsible for a depletion of the Alamazoatra and Alamaizina in the early nineteenth century.

## *Imperialism and Cattle*

Cattle-raising was the chief occupation of the pastoral Bara, Mahafaly, Antandroy, Sakalava, and Tsimihety, but imperial expansion from the mid-1790s resulted in the Merina elite capturing large herds. A small domestic

Figure 4i. Cattle grazing in the Ankay, 1862.
Source: Samuel Pasfield Oliver, *Madagascar and the Malagasy: With Sketches in the Provinces of Tamatave, Betanimena, and Ankova* (London: Day and Son, 1866), 40.

Figure4ii. Cattle pens in the *Alamazaotra* c.1880.
Source: James Sibree, *A Naturalist in Madagascar* (London: Seeley, 1915), 254.

market (cattle were considered sacred and meat consumed only at major festivals) was compensated for by vibrant Mascarene demand—for draught animals on the plantations, and meat to feed a burgeoning population (Mayeur 1785: 226; Campbell 1985: 297–9). As in Imerina cold winters and lack of fodder hindered development of pasture, and firing savannah to encourage fresh growth for pasture resulted in such erosion "as to leave many acres totally inaccessible to cattle" (Hastie 1824–25: 11; see also Mayeur 1777; Razafintsalama 1983: 195), cattle traders—chiefly the crown and court elite—sought land for pasture en-route to the coast. Initially, herds were kept in Antsihanaka and the Ankay, where they were sold directly to Mascarene traders (Mayeur 1777: 152, 186; Dumaine 1790: 260, 264, 268, 286), but following Merina military expansion to Tamatave in 1817, traders drove cattle directly to east coast ports. Thus, by the early 1800s, portions of the western fringe of the Alamazaotra had been cleared for cattle pens, as had small areas within the forest. Livestock grazing and trampling (on seedlings) disturbed stands and prevented normal regeneration. Extensive rearing, as in the Ankay, involved repeated firing—which destroyed degraded stands and affected intact forest edges (RPPD 2010: 32).

### *Construction*

In 1777, Mayeur commented that almost all plateau houses were earthen, or stone covered with earth and reeds. Few could command the labour necessary to fetch construction timber from the eastern forest. Thus, Jacques La Salle (late 1700s) claimed that only Andrianampoinimerina possessed a wooden house (Mayeur 1777: 174; Nativel 2005: 33–4; see also Belrose-Huygues 1975: 184). However, imperial expansion brought the Merina elite battle booty and thus the wealth to build wooden houses—authorised provided they were smaller than those of the sovereign (Callet 1974: 800; see also Ellis and Freeman 1842: 4).

Andrianampoinimerina's consolidation of political power in the 1790s was expressed symbolically through the building of new and bigger *lapa* (palaces)—ten at Ambohimanga, and nineteen at Antananarivo (Tacchi 1892: 485; Nativel 2005: 49–51). Mahandrihono, his principal Ambohimanga *lapa*, was constructed around an unprecedented 10m-high *andry* (central pillar – Wright 2007: 67). He also enlarged existing royal houses at Antananarivo (Mahitsy, Besakana, Masoandro), Ambohimanga (all bar one), Alasora (Irarisambo) and Ambohidratrimo (Imangabe). The dimensions of Manatsaralehibe, the largest *lapa*, are unknown but the Besakana was 7.05m x 6.39m (45m$^2$), with 3.17m high walls, while Mahitsy measured

9.61m x 8.07m (1.43m²) with 4.30m high walls (Le Sage. 1816: 61–2, 85; Ellis 1838. vol.2: 133, 135; Nativel 2005: 52).

From 1792, Andrianampoinimerina moved from Ambohimanga to Antananarivo which consequently experienced a major building boom. Until 1868, only three types of houses were permitted there. The first, built of *zozoro* (a tall sedge) split open, sun-dried and bound in 1.7 x 0.8m sheets to form walls that were strengthened by long bamboo pieces, and plastered with a waterproof mixture of clay, sand and cow dung. The second, *tery latsy* ("having bamboo on both sides"), were built of bamboos and small trees placed upright, strengthened by horizontal planks on both sides, and plastered. The third 'elite' house type used centrally-placed poles (large hardwood trunks), and *tandro-trano* ("house horns"—gables) which overlapped and projected out 0.5–0.6m—representing the horns of cattle (a major form of individual capital). Those with vertical plank walls were called *Kotona*; and *trano sarendry* (lit. "melted house") those in which the planks were laid at a slant of about 45 degrees. Chardenoux (1816) noted of the *Andriana* (noble caste) residences of Antananarivo:

> All their houses are built of wood bound by cords. They are attached to the ground only by 4 corner poles, which are very strong, and the 4 ridgepole

Figure 5i. *Trano Kotona*, Imerina c.1880.
Source: private collection.

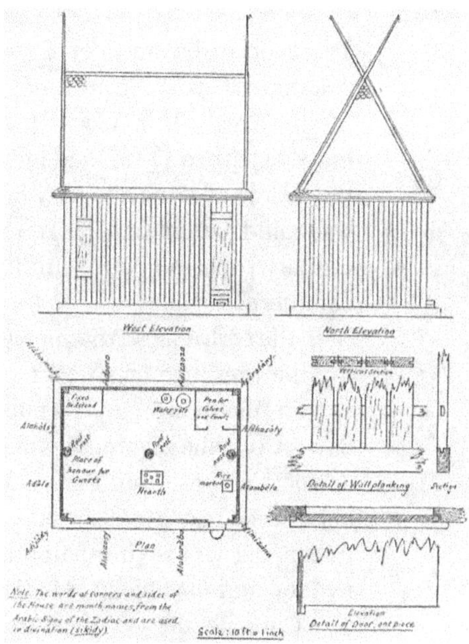


Figure 5ii. *Trano Kotona* house plan.
Source: Sibree, *Naturalist in Madagascar*, 96.

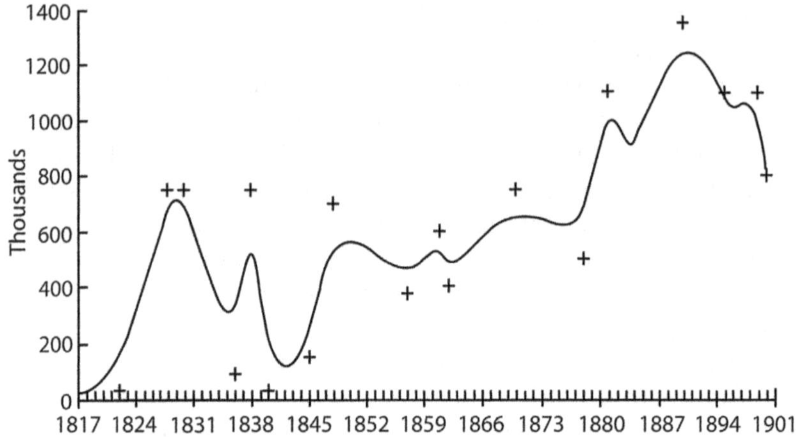


Figure 6. Antananarivo: population estimates, *1820-1901*.
Source: adapted from Campbell, *Economic History*, 139.

> forks which press them to the ground – which is very hard – and the only opening they have is a high and very narrow door, and a single window on the same side, so they are very dark [inside]. They are covered in reeds and very high (Chardenoux (1816) quoted in Belrose-Huygues 1975: 183–4; see also Sibree 1915: 184; Mason 1959: 6; Nativel 2005: 36 ).

There occurred another three main bursts of house construction in Antananarivo during the nineteenth century: following the 1820 Britanno-Merina Treaty; from 1829–34 when Antananarivo was transformed into a massive garrison town and industrial centre; and from 1862–68, following the death of Ranavalona, when Madagascar was re-opened to foreign trade and settlement. There was also, from 1820–53 onwards, sustained military expansion and with it the establishment of Merina garrisons in conquered regions.

In the early 1800s, Antananarivo's population fluctuated greatly, reflecting the presence or absence of large numbers of soldiers and workers. Hugon (1808) estimated its population at 25,000; Hastie (February 1821) at 8,000; Coppalle (September 1825) at 12,000 to 15,000, N. Leminier (1825) at 20,000; Carayon ( mid-1826) at 6,000 to 8,000; and Désiré Laverdant (1844) at between 15,000 and 20,000. Overall, it is probable that the population of Antananarivo rose from around 17,000 in the early 1820s to about 50,000 by 1860 and 100,000 by 1870 (Hugon 1808: 17; Jones 1820b; Leminier 1970: 795; Coppalle 1825–6 : 26; Carayon 1845: 9–10; Laverdant 1844: 24; Campbell 2005: 137). However, there was no necessary correlation between the city's population and house building which was particularly frenetic from 1829–31 when heavy military recruitment transformed Antananarivo into a major garrison town as young male recruits were obliged to attend Mahamasina, the military training ground at the capital. Their families often built residences there to house the draftee, his slave/s, and visiting relatives in-between military expeditions, which occurred each dry season and often entailed a 50 percent mortality rate. Regardless of population fluctuations, from 1829–31 Antananarivo experienced a construction boom, and demand for massive hardwood timber probably greater than in the years 1862–67 when there was a considerable influx of foreigners, but a marked drop in militarization (Freeman 1831; Johns 1832; Raombana 1853a: 247).

The traditional Merina house was rectangular, aligned north-south, one-roomed (with a loft in the northeast 'ancestral' corner), with a one-piece wooden door (at the southwest end), and one window (narrow, from 60cm to 1m above the earth—at the west or northwest end). The poor used *fazoro* rush windows; the wealthy, wooden shutters. By 1821, the average elite house was 3.05m x 7.32m (22.33m$^2$—the largest were 70m$^2$) and 5.5m high, "constructed of massive timber and well thatched with a triangular bulrush" (Hastie 1821). By contrast, ordinary earthern and reed houses were probably, like those of Betsileo: from 7.44m$^2$ to 10.03m$^2$ in area, with 0.91m to 1.52m high walls. The high pitched thatched roof was made of *herana* (*Cyperus latifolius*)—a strong sedge, different varieties of *horona* grass (e.g. *Aristida adscensionis, Pennisetum sp.*; *Rottbellia sp.*), or *mololo*—beaten rice straw. The houses of the elite had 2.5m to 4m long *tandro-trano*—some five to eight times longer than those of the 1790s (Sibree 1915: 182, 184, 236–7; Wills 1885: 127; Mason 1959: 7; Nativel 2005: 36).

From the 1820s, foreign influence inspired what Didier Nativel terms the "monumentalisation of royal buildings" (Nativel 2005: 73). Robert Farquhar, British governor of Mauritius, had instructed Hastie to demonstrate to the Merina the virtues European-style houses, but the only innovatory feature of his single-storey house at Antsahavola was a verandah and shuttered openings in the Mascarene Creole style (Coppalle 1825–6; Raombana 1853b: 129).[4] More influential was David Griffiths' Ambodin' Andohalo residence built chiefly by Louis Le Gros between January 1822 and December 1823 which, albeit thatched, had two storeys (two other two-storeyed buildings existed in the *rova* or royal enclosure) and encircling verandahs. It possessed the largest wooden frame ever seen in Madagascar and on completion was the biggest non-royal residence: It measured 20.73m x 6.71m (139m$^2$), less than the royal residence used by

[4] When in town, Hastie resided in the enclosure designed for the royal wives and concubines next to the Tranovola – Hilsenberg and Bojer, "A Sketch," 268.

Figure 7i. Ambodinandohalo, the house and missionary chapel.
Source: David Griffiths, *Hanes Madagascar* (Machynlleth: Richard Jones, 1843), i.

Figure 7ii. Tranovola.
Source: Ellis, *Three Visits to Madagascar,* 380.

David Jones as a school which measured 12.19m x 14.27m (173.95m$^2$), but some six times the ground floor area of the average house of the indigenous elite in Antananarivo (Hastie 1821; Belrose-Huygues 1975: 193; Ellis 1838. vol.2: 290).

The size of Griffiths' building is reflected in the cost of its construction. In 1829, the average value of the wealthiest Merina house was from $100 to $350, and of the *zozoro* house of the poorer inhabitants, only $3 to $4. By contrast, the average missionary house cost between $1,000 and $1,500 to build: Excluding the cost of Malagasy labour and timber ('donated' by the crown), Griffiths' property at Ambodin 'Andohalo was valued at $2,500; Johns' (house and yard) at $1,450; and Freeman's (house and yard) at $1,200 (Raombana 1853a: 399, 420–1, 423, 425–6; Lyall 1954: 230; Ellis 1838. vol.1: 95–6).

Ellis commented of Griffiths' building: "the frame… being the largest ever erected in the capital… excited much interest and surprise amongst the natives" (Ellis 1838. vol.2: 290–1). The wealthiest Merina hired Le Gros to build similar buildings. Radama, making his innovatory mark on royal architecture, had Le Gros transform his Marivolanitra residence into a two-storey building in 1822, and construct the Tranovola, completed in

1825, as a square, two-storeyed building, "surrounded by a kind of wooden balcony, very rudely carved, to which an ascent is obtained by a ladder, for want of a better stair-case" (Hilsenberg and Bojer 1833: 267), with a high-pitched shingle roof extended to overhang the balcony. Le Gros also built Soanieranana Palace of which Carayon noted in mid-1826, when the building was still far from completion, that it then comprised "a central single-storey residential block, with two wings on each facade, all forming a square, each side of which measured thirty-six metres" (Carayon 1845: 9; see also Raombana 1953b: 6; Grandidier 1908: 537; Nativel 2005:75; Belrose-Huygues 1975: 192, 196–7; Copland 1822: 81; Coppalle 1825–6).

Figure 8. Soanieranana Palace.
Source: E. Colin & P. Suau, *Madagascar et la mission catholique* (Paris: Sanard et Derangeon), 20.

Figure 9. Antananarivo, 1856.
Source: Ellis, *Three Visits*, frontispiece.

Carayon commented that only Radama's palace, Griffiths' residence, and "small houses in European style" built by the missionaries, distinguished Antananarivo from other plateau settlements (Carayon 1845: 8–9). Viewed from the east thirty years later, one can discern nine two-storeyed houses, possibly with verandahs, that, excepting the roofs of some—of shingle in the Mascarene style—resemble Le Gros' plan for Griffiths' residence (Callet 1978: 1066; Ellis 1838. vol.1: 95–6; Dez 1970: 23). Moreover, the capital's influence spread. Ellis (1856) noted an exception to the otherwise traditional houses of Ambatomanga:

> It might be about sixty feet long, and half as wide. It was two stories high, with door and windows in each story, and a steep roof with attic windows in the sides and the ends. The side walls were sheltered by two verandahs one above the other, and the posts which supported the two verandahs were upwards of twenty feet high. This remarkable building, with walls of wood framed in diagonal panels, roofed with shingles, spacious, more than

Figure 10i. House of the chief, Ambatomanga.
Source: James Sibree, Madagascar et ses habitants. Journal d'un séjour de quatre ans dans l'ile. trans. H. Monod et Henri Monod (Toulouse : Société des Livres Religieux, 1873), 89.

Figure 10ii. Large house, Lazaina.
Source: William Ellis, The Martyr Church (London: John Snow, 1870), 350.

> double the height of any other houses, and European and attractive in its form, was standing upon the highest spot in the village, surrounded by a wall (Ellis 1859: 336–7).

In Lazaina, Ellis found another large two-storeyed house with a verandah, though with a thatched rather than shingle roof. European styles in furniture were also adopted. For instance, whereas traditionally the Malagasy slept on mats spread on the floor or (the *Andriana* or noble caste) on raised platforms, from the 1820s wealthier Malagasy adopted European-style wooden beds (Nativel 2005: 50).

There was also an expansion in the number of Merina garrisons and associated settlements of Merina colonists. These made extensive use of wood. As LMS missionary Richard Baron stated: "all the towns and villages with Hova [Merina] Governors are surrounded by palisades, frequently in a double series, made of the trunks of young trees, six or eight inches in diameter, fixed in the ground and placed in contact with each other. I once counted the trees that had been thus used in a certain village, and found that there were about 10,000. These trees, moreover, in many of these places are renewed every eight or ten years...Even when stone and lime or other suitable materials are abundant and close at hand..." (Baron 1891: 324).

As almost all buildings had earth foundations, the most non-perishable hardwoods were sought as corner posts and central columns. For example, Merana (*Vernonia merana*) wood was chosen for the posts in Andrianampoinimerina's Manatsaralehibe palace (Tacchi 1892: 495). The most valued woods and their uses are shown in the following table:

Table 1. Woods used notably in house construction in Imerina[1]

| *Name* | *Qualities* |
|---|---|
| *ambora* (*Tambourissa parvifolia*, Baker) a term used to cover several species | used for the frames of doors and windows in houses of nobles and sovereigns. It was used to construct the balconies of the 'Silver Palace'. Also used for making coffins. |
| *amborasevoka* (Tambourissa sp.) | |
| *fantsikahitra* (*Plectronia sp.*) | hardwood |
| *harahara* (*Phylloxylon decipiens*), | an extremely hard tight-grained wood used to make amongst other objects, spear shafts, spade handles and musical instruments |

(*Continued*)

Table 1. (*Cont.*)

| *Name* | *Qualities* |
|---|---|
| *hazo tokana* ('lone tree' – *Vernonia sp*) | hard durable wood used to make corner poles and pillars – but it was believed that if used alone it would cause pottery to break |
| *hazomena* ('red wood' – *Ochna madagascariensis or einmannia rutenbergii*, Engler) | a hard red wood that could reach 60 metres in height; and which lasts when planted in the soil – excellent for corner posts for which much used; and pillars, and partition posts planted in the ground. It was also used in the construction of the 'Silver Palace' |
| *hazondrano* (*Ilex mitis*) | a hard wood from the straight-trunked evergreen tree, between 10 and 25 m tall, used in the manufacture of poles of men's palanquins |
| *hazotokana / mamanotsa*, (*Vernonia sp*): related to the *merana* | a hard wood |
| *hetatra* (*Podocarpus madagascariensis or Podocarpus Thunbergii*, Hook) a tree similar to the yew and a species of conifer (Podocarpaceae family), | a hard white wood used to make beams, and doors (not single-piece), floor boards, planks, and joists; in the construction of palanquins, and in ordinary carpentry |
| *keliravina* (*Breonia keliravina*) | |
| *lalona* (*Weinmannia Bojeriana*, Tul.) | used particularly in making partitions, joists, and door and window frames. |
| *merana* (*Vernonia merana*, Baker), related to the *Vernonia sp.* | a rot-proof hard wood used to make corner poles and pillars. |
| *nato/ nanto* ('hard' – *Labramia bojeri* [Beach Apricot Tree]. | hardwood used by Betsimisaraka in making canoes and coffins |
| *sary* | a white wood used in house building |
| *tandrokondry* | |

Table 1. (*Cont.*)

| *Name* | *Qualities* |
|---|---|
| *valanirana* (*Nuxia capitata,* Baker) | hardwood used in house construction; and one of the woods used as firewood for the monarch at the *Fandroana* or New Year. |
| *varongy* (*Ocotea tricoplebia*) | used chiefly in beams and to make single-piece doors; also, roof joists, floors (and rice mortars)—notably the varieties *varongifotsy*; *varongiimainty* (*Dracaena reflexa*), the only species of *varongy* Richardson mentions as being used for the building both of houses and canoes; *varongimena,* and *varongira 'vimbanana.* |
| *voamboana* (*Dalbergia Baroni,* Baker) also known as 'Palisander Rosewood' or 'Madagascar Rosewood') | a very light pleasant smelling and valued wood much used in for furniture, doors, floors and general house construction |
| *valamia* [?]- | large tree used to make planks, and the gable poles for Merina houses |
| *volombodipona* (*Dalbergia purpurescense,* Baillon) | a hard dark-red tree, used to make the floor of the 'Silver Palace,' and walking sticks. |

[1] HdR vol.1, 58; see also ; J. Richardson, *A New Malagasy-English Dictionary* (Antananarivo: LMS, 1885), 32, 167, 231, 203, 244–5, 255, 280, 372, 424, 438, 442, 446, 448, 556, 730; 740, 759, 765, 768, 778, 944; Philippe Beaujard, 'Plantes et médicine traditionnelle dans le sud-est de Madagascar' *Journal of Ethnopharmacology,* 23 (1988), 205; Sylvain G. Razafimandimbison, 'A Systematic Revision of "Breonia" (Rubiaceae-Naucleeae)' *Annals of the Missouri Botanical Garden,* 89. 1 (2002), 3; http://zipcodezoo.com/Plants/L/Labramia_bojeri/ (26/05/10); [James] Wills, 'Native Products used in Malagasy Indsuries' *AAMM* 9 (1885), 125; E. Drake del Castille, 'Botanique' in Raphaël Blanchard (ed.), *Madagascar au début du XXe siècle* (Paris: Société d'éditions scientifiques et littéraires, 1902), 120–3.; L. Carayon, *Histoire de l'établissement français de Madagascar pendant la Restauration précédée d'une description de cette ile et suivie de quelques considérations politiques et commerciales sur l'expédition et la colonisation de Madagascar* (Paris : Gide, 1845), 191–2; Nativel, *Maisons royales, demeures des grands à Madagascar* (Paris: Karthala, 2005), 34–6; V.A. Barbié du Bocage, 'Ambassade Anglaise à Tananarivou (Madagascar)' *Bulletin de la Société de Géographie* série 5, tome3 (1862), 174 ; W. Deans Cowan, *The Tanala* (Faravohitra: FFMA, 1881), 12–13; James Sibree, *A Naturalist in Madagascar* (London: Seeley, 1915), 158–9.

However, certain otherwise eligible woods were avoided because of their allegedly sinister effects:

Table 2. Woods not to be used in construction in Imerina[1]

| *Name* | *Reason* |
| --- | --- |
| *alimpangady* ['night spade?'] | a wood reserved for the dead |
| *amontana* – a large tree with soft wood, of which Richardson notes two types: *Ficus Baroni*, Baker and *Ficus trichophlebia*, Baker. | attracts lightening |
| *andrezina* (*Trema grisea*) | believed to attract lightening |
| *aviavy* (*Ficus sp.* Moracees), Fig tree. | its roots were believed to suffocate other trees and therefore it was believed to cause the death of those who transported it, and endanger the proprietor of the house in the construction of which it is used. |
| *foraha* (*Calophyllum inophyllum*), a tall tree that grew in the sandy coastal soil | the wood of which could be used in boat construction but not in house construction lest it attract lightening. |
| *harongana* (*Harongana madagascariensis*) | could prove fatal to even a healthy house-owner |
| *hazomafana* ('warm wood') | kills young those who transport it |
| *havozo* [*Agatophyllum aromatica* – ex *Ravensara aromatic*?] | makes one debile. |
| *lambinana* (*Nuxia sphaerocephala*, Baker) | associated with the dead – used to make the stretchers for transporting the dead. However, Richardson notes that it was used in house construction. |
| *tambintsy* (*Psorospermum lanceolatum* (*Choisy*). Hochr. | diminishes and destroys descendants |
| *tandrano* | |
| *Vintana/ vintanina* ('exorcism'), or *Karazan-kazo* (*Calophyllum spurium*) – a type of tacamaca | a straight tall tree, that grew in the marshes and mountains, widely used in canoe construction but |

Table 2. (*Cont.*)

| *Name* | *Reason* |
|---|---|
| wood.red and white; | avoided in house construction because it was believed to attract lightening and makes one die young |
| *voara* (*Ficus tiliifolia*, Baker. *Moraceae*). | would badly affect the proprietor of the house (though can be used to make rice pans) |

[1] HdR vol.1, 58 ; Richardson, Dictionary, 33; Richardson, Dictionary, 374; Philippe Beaujard, 'Plantes et médicine traditionnelle dans le sud-est de Madagascar' *Journal of Ethnopharmacology,* 23 (1988), 189, 192, 256 ; Nativel, *Maisons royales, demeures des grands à Madagascar* (Paris: Karthala, 2005), 34–6; Marion Viano, 'Rôle des pratiques paysannes Betsileo sur la dispersion des graines par les oiseaux en lisière du corridor forestier de Fianarantsoa (Madagascar)' Mémoire DEA, Université d'Orléans (2004), annexe 6 ; Gillian Feeley-Harnik, 'The Sakalava House' *Anthropos* 75 (1980), 564 ; 'Madagascar and Malagasy' http://www.malagasyencyclopedia.com/bins/teny2 (26/05/10); See. http://www.guerissure.ch/ravensares_ravintsares_agatophylles_et_affilies.html (26/05/10).

Figure 11i. Woodcutter's village, near Alamazaotra.
Source: William Ellis, *History of Madagascar* (London: Fisher, Son & Co.) vol.1, 35.

Figure 11ii. Evidence of felling in Ankay section of the Alamazaotra, 1862.
Source: Vincent W. Ryan, *Mauritius and Madagascar. Journals of An Eight Years' Residence in the Diocese of Mauritius, and of a Visit to Madagascar* (London: Sheeley, Jackon, and Halliday, 1864), 286.

As house-size increased, so did the trunks used as pillars and beams—and the labour to obtain them. Farquhar (1817) had advocated forming "bands of foresters and woodcutters" (Farquhar (1817) in Ellis 1838. vol.2: 164) and the Merina crown quickly established permanent woodcutter *fanompoana* units (termed the *foloroazato* – lit. "the 1,200") whose members grew from 1,200 to 2,000 by the 1880s (Oliver 1886. vol.2: 89, 192–3; Grandidier 1908: 370).[5] To drag an average tree trunk to Antananarivo required 75 men. A 5,000-strong *fanompoana* unit was drafted to carry a 39m-high trunk for the Manjakamiadana palace. Raombana noted of royal constructions: "Their pillars are enormous and has excited the astonishments of Europeans who has seen them... The getting and fetching of the enormous pillars and other Timbers for the construction of these Royal Houses fatigued the people very much" (Raombana 1980: 232–3; see also ibid, 129; Callet 1974: 57; Hewlett 1887: 296; Chapus 1925: 206; Freeman and Johns 1840: 44; Grandidier 1916: 29–40).

The Alamazoatra supplied timber for Griffiths' house—a clear indication that by 1822 the Alamaizina had no large hardwoods (Griffiths 1843: 30; Ellis 1838. vol.2: 290–1, 309). By 1839, insufficient wood being found in the western fringe of the Alamazoatra to help build a royal house, Ranavalona ordered *fanompoana* "to fetch in (branches) of the trees planted in the gardens, and hedges, or fences..." (Rainitsontsoraka in Ellis 1857: 69) and construction timber was sought increasingly within the forest—of which it was commented in 1841: "Many are engaged, going and returning, in conveying timber from the forest, for building houses for the queen" ([four Malagasy Christians] (1841). In [J.L.K] [1849]: 147). In 1862, Ryan observed a wood-cutter's hut at Alamazoatra, west of which the dense forest ended (Ryan 1864: 286–8, 290). By 1867, construction wood was being sought in the north-eastern Alamazoatra by Sihanaka *fanompoana*, or its southern sectors by the Betsileo or Tanala. The Ankay section provided only rounded wood [*boribory*?] and the *hafotra*, the fibres of which were used to tie pieces of wood together (Callet 1974: 57; Hanson 2008: 269).

Wastage accentuated the problem. Although Merina carpenters (Andriana Andriantompkoindrindra and Andrianamboninolona) were reputedly the best in Madagascar, they traditionally axed tree trunks into two planks and discarded the rest, and most doors were one-piece units fashioned from single trunks. Although by the 1830s, many had adopted the pit-saw, and were thus much more efficient, many persisted with traditional wasteful methods (Callet 1974: 596; Anon [c.1790]: 60; Mayeur 1777: 174; Chapelier 1905–6: 16, 23; Raombana 1953b: 6; Oliver 1886. vol.2: 83, 91, 96–7; Grandidier 1908: 369 and 1928: 104, 133; Couland 1973a: 137–42;

[5] Governors of Fianarantsoa, the chief Merina garrison in Betsileo, maintained a permanent *fanompoana* of 900 men, including 700 woodcutters and 100 carpenters – Nilsen [1943].

Wills 1885: 125; Nativel 2005: 34, 52: Ellis 1838. vol.1: 317–8). In addition, fires were a constant hazard in concentrated settlements built of combustible materials. For example, major fires destroyed the main palace in Antananarivo (1815); Le Gros' Andohalo workshop and fifteen neighbouring houses (1826); and Camerons's Ambatonakanga workshop along with many neighbouring residences (1828/9). Again, in mid-1826 the gunpowder store at [Masoarivo], 5km southeast of Antananarivo, exploded, destroying Radama's half-built residence there (Dez 1970: 14; Le Sage 1816: 61–2, 85; Ellis 1838. vol.2: 133, 135; Coppalle 1825–6: 49).

In all, the construction boom in Antananarivo from the 1820s, and the increase in the size of the houses of the elite, combined with wasteful forms of cutting wood, played a significant role in the depletion of the forests to the east of the capital.

### *Firewood*

The growth of Antananarivo, and the often bitterly cold winter months between April and October, ensured a heavy demand for firewood. Ordinary people scavenged the rapidly decreasing patches of hilltop forest. Andrianampoinimerina banned exploitation of Ambohimanga forest except (upon royal authorization) for firewood to warm Andriana wives during birth-giving (Callet 1974: 575–6). He also preserved the fruit-bearing trees (notably the *voankandrina* (*Garcinia gerrardi.* Baker) south of Ambohijanhary, a measure upheld by Radama but rescinded by Ranavalona who permitted the collection of firewood there (Callet 1974: 611). Nevertheless, as early as 1777, Mayeur observed that firewood was so scarce in Imerina that people substituted grass, while Charles Campbell (1840) observed of the Ambatomanga region: "the inhabitants have no firewood within 15 miles and are obliged to cook their food with grass" (Campbell 1963: 472–3; see also Tourte 2003: 98). To supply court needs, Sihanaka *fanompoana* collected firewood from the northeastern forest (Campbell 1980: 346) – in the process contributing to its depletion.

### *Autarky and Industry*

As detailed elsewhere, by the mid-1820s, Radama judged the 1820 Britanno-Merina Treaty to have failed, and consequently adopted an autarkic policy aimed at generating economic development and imperial expansion through exploiting domestic resources and the foreign personnel and skills already present in Imerina. However, autarkic policies accelerated depletion of forest resources (Campbell 1987: 395–411).

LMS artisan John Canham, who arrived in 1822, established a workshop teaching tanning and shoe-making to crown-appointed apprentices. Because of the indigenous habit of eating meat with the hide attached, Canham at first imported leather from Mauritius, but Radama issued orders that the hides of all slaughtered cattle be sent to Canham who thus obtained over 16,500 hides in 1828 alone. Initially prevented from tanning by the lack of lime, Canham used bark from the *lalona* (*Weinmannia bojeriana*) from the eastern forest and *nanto*, a red mahogany-like hardwood from the northeastern forest. Later, private tanneries in Antananarivo used *paiso* or peach tree (*Prunus persica*) bark (Cameron 1874: 7; Griffiths 1843: 38; Oliver 1886. vol.2: 84–5; Grandidier 1928: 133–4; Hastie 1817: 128; Raombana 1853b: 16–17; Callet 1974: 927). Under Ranavalona, Canham relocated to Anosifisaka, where his tanning industry, fed by lime deposits discovered on the Sirabe plain 70 km south of Antananarivo, and by *fanompoana* labour, resulted in such population growth that its name was changed to Ambohimandroso ("Village of Progress"). He established two similar tanneries at Andoharano and Vodivato. All produced military boots, belts, saddles and pouches.[6] After Canham left Madagascar in 1834, tanning was maintained by the government, chiefly at Sodivato, to meet military needs (Wills 1885: 123; see also Campbell 2005: 95, 247; Chauvin 1939: 19).

The iron industry had a far greater impact on forest resources. Imerina was rich in iron ore deposits, exploited from the mid-sixteenth century to produce chiefly agricultural and wood-cutting tools, and weapons. By the late 1700s, diminishing woodland had driven the industry to Amoronkay, on the western border of the Analamaizina forest. Indeed, 'Amoronkay' derives from 'Tan-kay' ("the country that burns")—the *Tantara* (traditional history of the sovereigns of Imerina) recording that "the entire frontier of the forest was burnt" (Callet 1974: 70; see also Ellis 1838. vol.1: 307–9; Oliver 1886. vol.1: 343, 488, 492 and vol.2: 88–9; Raombana 1853b: 15; Grandidier 1928: 179, 181; Mayeur 1775: 36 and 1785: 210; Hastie 1817: 169–70, 182–3; Sémerville 1933: 28; *Decary* 1966. vol.1: 29; Ralalarimanga 2010: fn.4).

---

[6] Droit arrived in Antananarivo from Majunga about 1832, and Laborde shortly afterwards from Mananjary, after he had been shipwrecked on the south east coast of Madagascar en-route to Europe from India – Raombana 1853a: 307–8; Campbell 1988.

This was probably the region visited by Hugon in 1808. A little west of 'Voni-biamang' [Ambatomanga?], on the Merina-Ankay frontier, he noted charcoal works at 'Voni-dzahare' [Ambohizanahary'?] from where the Merina also obtained construction wood. The nearby village of 'Toa-bouitche' ['Toabotsy'?] was inhabited by 'Zafi-mamé' [the Zafimamy of Alahamadintany in northern Imerina?—he later appears to refer to them as 'Zafimanif'), full-time wood-cutters (making planks or beams) and charcoal burners. They sold their products in 'd'cessasse' [?] market to traders from Central Imerina in exchange for provisions (including manioc and meat). The 'Zafi-mamé' may have constituted *fanompoana*, as from the 1820s, the government maintained a permanent *fanompoana* unit of 500 to 600 charcoal burners; 400 smiths and general iron workers; as well as smaller numbers of gunsmiths and spear makers. Oliver comments that Ellis' observation that these units were "a sort of Government slaves," confined for life to their villages and occupations, were as valid for the 1880s as when made in the 1830s (Oliver 1886. vol.2: 89, 192–3; see also Grandidier 1908: 370; Hugon 1808: 13–13b).

In 1785, Mayeur noted frequent 1.8kg to 2.3 kg surface iron ore deposits in Amoronkay which were worked in 1m$^2$ to 1.6m$^2$ pits to produce two types of smelted ore: *osa* ("soft iron") used to manufacture spades, square nails and hinges; and steel, to make hatchets, knives, scissors, saws, and hammers. Indigenous smiths also produced high quality anvils, hammers, chisels, planes, scissors, knives, needles, awls, tripods, scales, railings, slave irons and chains, pots, bayonets, spearheads and musket balls; while Merina Andrianando smiths specialised in repairing foreign musketry (Mayeur 1785: 210, 228; Idem 1777: 165, 171, 175; Idem 1775: 36; Hastie 1817: 169–70, 182–3; Hilsenberg and Bojer 1833: 263; Sémerville 1933: 28; Ellis 1838. vol.1: 307–9; Oliver 1886. vol.1: 343, 488, 492 and vol.2: 88–9, 90–1; Raombana 1853b: 15; Grandidier 1908: 185, 313 and 1928: 179, 181, 185, 216, 294; Decary 1966. vol.1: 22–7, 29, 31–2, 37, 49, 53–5; Anon [c.1790]: 58–9; Lescalier 1792: 327; Chapelier 1816): 174; Chapus and Dandouau 1951–52: 64; Coppalle 1825–6: 45–7; Callet 1974: 70).

Mayeur commented on the site:

> almost all were engaged in work. Some open up iron ore mines, others construct furnaces of sandstone, built on good foundations and faced with clay, two feet in diameter and four feet deep. [There exist] two bellows to fan the charcoal fire that has been lit below. The Merina start by placing a thick bed of charcoal on the bottom of the furnace, followed by a thinner layer of well-washed iron ore, upon which another layer of charcoal is laid and so on until it is full. They then work the bellows, continuously and energetically,

Figure 12. Path in the Analamaizina, near Andrangaloaka, c.1865–85.
Source: John Parret-USC Digital library, impa-m28 – http://digitallibrary.usc.edu/search/controller/view/impa-m28.html (17/05/10).

Figure 13i. Smelting iron ore, Amoronkay, early 19th century.
Source: Ellis *History of Madagascar* vol.1, 308.

Figure 13ii. Smithy c.1910.
Source: Postcard, private collection.

> until the ore has melted. This procedure lasts about four or five hours and regularly produces approximately forty to fifty pounds of iron. Once the mass has cooled sufficiently, they select the purest iron for forging; it is of high quality. The rest, they pour back into the crucible a second time in order to fully eradicate the impurities (Mayeur 1785: 210; see also Idem 1777: 174; Lescalier 1792: 327–8; Oliver 1886. vol.2: 88–9; Grandidier 1928: 179).

In 1817, Hastie noted one site comprising two workshops containing 18 and 32 forges respectively. The Malagasy used the Southeast Asian double feather-piston bellows linked beneath by a tube, rather than African bellows. Five men and two boy apprentices produced on average 18kg to 25kg of iron per session, although up to 408kg has been recorded. Traditional methods permitted 40 per cent of worked iron to be transformed into pure iron, although this figure was based on only 25 per cent of the original mineral, the rest being lost during the smelting process. George Chick, the LMS blacksmith, claimed that Malagasy iron was as malleable as copper, and superior in quality to that of Europe (Hastie 1817: 169–70, 181–2; Grandidier 1908: 368–9 and 1928: 182–3; Sibree 1915: 156; Mayeur 1777: 174; Coppalle 1825–6 : 47).

Some workshops combined all the processes of production; Thus the five men and two boys noted by Hastie "can convert ore into 48 pounds [21.8 kg.] of iron and the iron into eight spades per day." Each spade weighed 2.27kg, was 150mm wide, 225mm long, and was completed by a 3.3m long handle of red hardwood. Smelted iron was also transported to market to be purchased by smiths who and worked from home (Campbell 2005: 36; Vassière 1885: 29).

Both processes involved considerable charcoal consumption. It took 7 to 8 hours to smelt 170 kg to 180 kg or iron ore, while in smithies, charcoal was laid on the open ground, occasioning considerable heat loss and thus fuel wastage. From 1820, as the Analamaizina was depleted of large 'construction' trunks, it became valued almost exclusively for firewood and wood for charcoal. Even that took its toll, and by the late 1800s the centre of iron production had moved from south of Mantasoa to Marorangotra, 8 km to its northeast (Raombana 1953a: 15-16 ; Couland 1973b: 238 fn.1, 241 fn.1).

Iron-working expanded dramatically with the influx of foreign craftsmen working under royal contracts. After his arrival in mid-1822, Chick instructed several hundred crown-selected apprentices at his metal workshop in Amparibe, near Antananarivo. Competition with local ironworkers obliged him to concentrate on the local elite market for ornamental ironwork and utilitarian novelties such as screws, locks, rounded nails and wire. However, from 1827 Chick made considerable losses until in early 1829 he concluded a royal contract to provide ironwork for Soanierana palace, sword and bayonets (from 1832–5 Amapribe produced possibly 48,000 quality bayonets and swords annually), and boilers for the armaments factory run by fellow LMS artisan James Cameron (Hastie 1817: 182; Campbell 1988; Ellis 1838 vol.1: 311–3; Cameron 1874: 6; Lyall 1954: 137–41, 171; Griffiths 1843: 15, 52; Oliver 1886. vol.2: 91; Campbell 1987: 403, 410; Chapus 1925: 80).

Indeed, it was Cameron who founded the first large-scale industry in Madagascar. Upon arrival in 1826 he instructed 600 apprentices in a large woodwork shop at Ambatonakanga, then northwest of Antananarivo. In 1827, Radama contracted Cameron to build a water mill (supplied by canal from the river Ikopa) to power a planned textile factory at Isoraka. However, Radama died in 1828 and, following the 1829 French assault on north-eastern ports, Ranavalona substituted the idea of an armaments for a textile factory— contracting Cameron to build it. As in 1829, his Ambatonakanga premises burned down, Cameron re-located to Analakely ("little wood"—possibly because there was a small wood there), a short distance northwest of Ambatonakanga, where he also produced soap. By 1835, his Isoraka gunpowder factory was described as being on an immense scale—earning him royal approbation and considerable wealth (Tacchi 1892: 489; Callet 1978: 1113; Ellis 1838. vol.I: 316–9, 321; Cameron 1874: 5; Toy 1875: 48–53; Sibree 1923: 314–6; Jordaan 1969: 1–11, 130; Campbell 1985: 248 and 2005: 92–3). Both production centres demanded considerable forest resources in the form of hardwoods (construction); wormwood

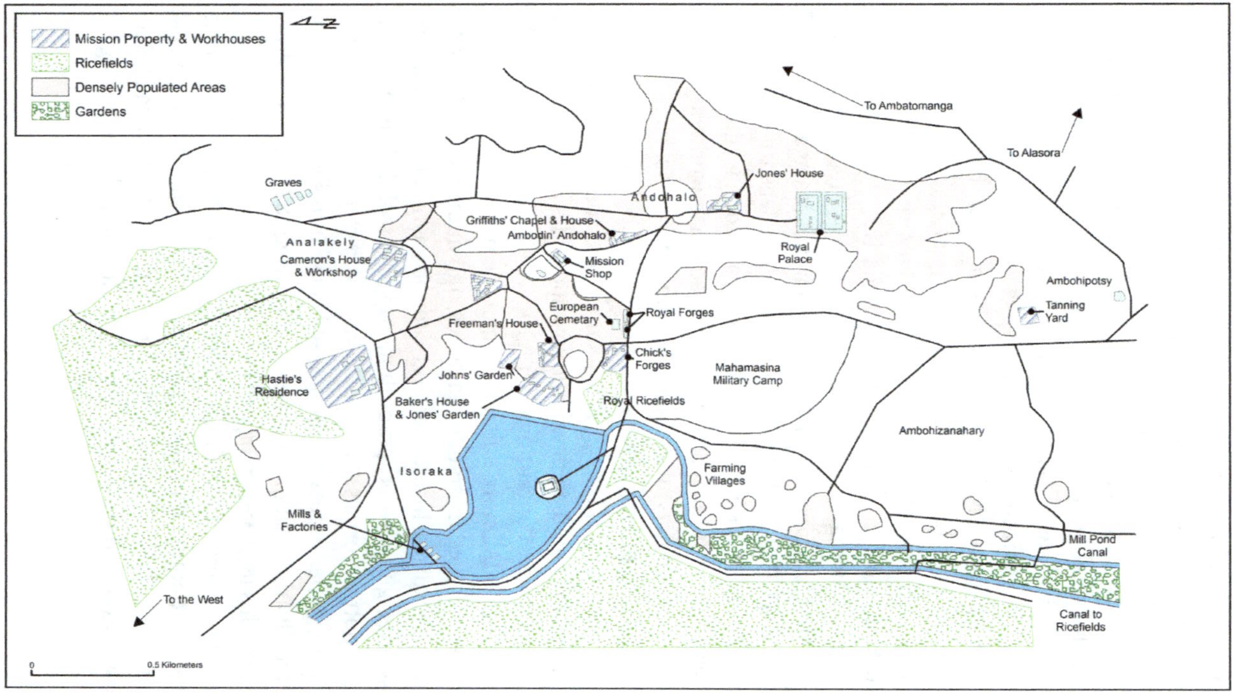


Map 2. Antananarivo: mission property and workshops, 1828-35.
Source: adapted from Gwyn Campbell, An *Economic History of Imperial Madagascar, 1750–1895* (Cambridge: Cambridge University Press, 1985), 94.

(soap manufacture); *boribory* (from which potash extract was used in both gunpowder and soap making); and wood (charcoal fuel) (Raombana 1853a: 357–60; Grandidier 1916: 39–40; Chauvin 1939: 8–9; Pfieffer 1861: 267; Raminadrasoa and Ratsivalaka 1979: 9).

A musket factory and cannon foundry were also founded at Andriambe, near Ilafy, that under Jean Laborde (1805–78) succeeded in "manufacturing exceptionally good gunpowder and muskets." at one stage at a rate of one a day, as well as a cannon, and bullets cast from an impure lead extracted from indigenous galena (Chapus 1925: 203).[7] With the 1835 ban on Christianity, the LMS artisans left Madagascar, but their workshops were taken over by trained indigenous personnel. The ban did not affect Laborde, who in 1837, due to the lack of fuel and water, relocated to Mantasoa, 40 km east of Antananarivo, situated close to both lakes and the Alamaizina. There, following Roret's technical manuals, he constructed "a true industrial centre, a kind of Malagasy Creusot" (Chapus 1925: 204). Although the workshops (one measured 54.86m x 11m—603.46m$^2$—and four were half that size) had stone walls, the complex used large amounts of construction wood, that workers (mostly slaves or Christian convicts) travelled 11km to 13 km to locate. An aqueduct carried water to the site where, in June 1843, the first blast furnace was lit; thirteen months later the first cannon was completed. In 1849, Laborde planned for nineteen new buildings and, at its height, Mantasoa comprised five factories containing blast furnaces—that smelt iron ore in far larger quantities and more efficiently than the traditional small scale bellow furnaces—forges, hydraulic machines, and numerous craft shops. The complex specialised in musket and cannon manufacture, but also produced swords, gunpowder, grapeshot, copper, steel, lightening conductors, glass, pottery, bricks, tiles, silk, a variety of cloths, candles, lime, dye, white soap, paper potassium, processed sugar into sweets and alcohol, and tanned leather (Chapus 1925 : 203–5; Hewlett 1887: 379–80; Oliver 1886. vol.2: 107; Ellis 1870: 176–7, 187, 204).

Under autarky, some processing of mineral and agricultural products also occurred in conquered lowland provinces. For example, on the east coast from 1840–5, ironsmiths under the direction of de Lastelle manufactured pickaxes, axes and shovels for export to Réunion, thus realising an

[7] Previously, Laborde, a wheelwright from Auch, in Southwestern France, had spent three years trading in India – Chauvin 1939; Raombana 1853a: 307–8.

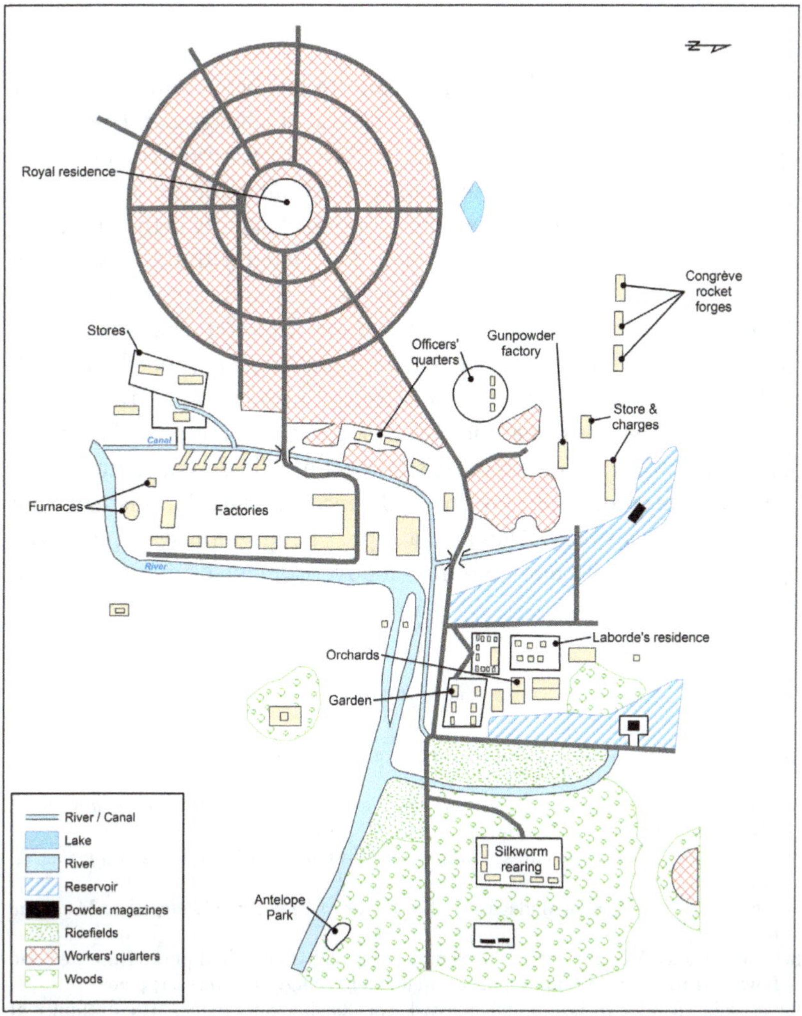


Map 3. Mantasoa industrial complex, c.1850.
Source: adapted from Campbell, *Economic History*, 98.

enterprise first conceived by Bourdonnais in the mid-eighteenth century. De Lastelle also established a shipyard at Mahela which employed 150 Malagasy carpenters, ironsmiths and coopers building and repairing boats employed on inland waters and on the coast (Campbell 1985: 297; Fontoynont and Nicol 1940: 25).

## *Conclusion*

Conventionally, *tavy* or swidden agriculture has been blamed for most historical destruction of the eastern rainforest of Madagascar. This article argues that, at least for the Antananarivo-Tamatave corridor, major depletion of forest resources occurred mainly as a result of Merina state policy, notably imperial expansion from the adoption of autarky in the mid-1820s and the drive for industrial production. Associated with this was a huge expansion in cattle grazing on the forest borders, and exploitation of forest resources: for hardwoods (construction), firewood, charcoal (industrial fuel), barks (lime), wormwood (soap), and *boribory* (potash). Indeed, exploitation of forest resources was central to imperial Merina policy. Hence, while there is some evidence to support conventional claims that the Merina crown preserved the eastern sector of the Alamazoatra (lower band of eastern forest) as part of its defense strategy, the western zone of the Alamazoatra and Alamaizina (upper band of forest) experienced significant shrinkage and resource impoverishment during the period 1790–61.

## *Bibliography*

Anon [c.1790]. "Mémoire historique et politique sur l'Isle de Madagascar" [Bourbon]. British Library. Add.18126.

Aubert, B. Razafiarison, S., Bertrand, A. eds. 2003. *Déforestation et systèmes agraires à Madagascar. Les dynamiques des tavy sur la côte orientale*. Montpellier: Cirad.

Baron, R. 1890. "A Malagasy Forest." *Antananarivo Annual and Madagascar Magazine*, 14: 196–211.

Baron, R. 1891. "The Flora of Madagascar" *Antananarivo Annual and Madagascar Magazine* 15:, 322–357.

Belrose-Huygues, Vincent. 1975. "Un exemple de syncrétisme esthétique au XIXe siècle: le Rova de Tananarive d'Andrianjaka à Radama Ier." *Omaly sy Anio* 1–2: 173–207.

Bond, W.J., Silander Jr, J.A., Ranaivonasy, J. and Ratsirarson, J. 2008. "The antiquity of Madagascar's grasslands and the rise of $C_4$ grassy biomes." *Journal of Biogeography* 35: 1743–1758.

Callet, R.P. ed. 1974. *Histoire des Rois*. vols 1–4. Tananarive: Librarie de Madagascar.

Callet, R.P. ed. 1978. *Histoire des Rois*. vol.5. Antananarivo: Académie Malgache.

Cameron, J. 1874. *Recollections of Mission Life in Madagascar during the early days of the LMS Mission*. Antananarivo: Abraham Kingdon.

Campbell, C.M. 1963. "Journal (1840)." *Studia* vol. 2.

Campbell, G. 1980. "Labour and the Transport Problem in Imperial Madagascar, 1810–1895" *Journal of African History* 21.3: 341–356.

Campbell, G. 1985. "The Role of the London Missionary Society in the Rise of the Merina Empire, 1810–1861." PhD. University of Wales.

Campbell, G. 1987. "The Adoption of Autarky in Imperial Madagascar, 1820–1835," *Journal of African History* 28.3: 395–411.

Campbell, G. 1988. "Slavery and Fanompoana: The Structure of Forced Labour in Imerina (Madagascar), 1790–1861" *Journal of African History* 29.3: 463–486.

Campbell, G. 2005. *An Economic History of Imperial Madagascar, 1750–1895*. Cambridge: Cambridge University Press.
Carayon, L. 1845. *Histoire de l'établissement français de Madagascar pendant la Restauration.* Paris: Gide.
Chapelier, L. 1905–6. "lettres adressées au citoyen préfet de l'ile de France, de décembre 1803 en mai 1805." Ed. Jully, Anthony. *Bulletin de l'Académie Malgache* 4 (1905–6): 1–56.
Chapelier, L. 1940. "manuscrits." Ed. Henri Poisson. *Collection de documents concernant Madagascar et les pays voisins* vol.2. Tananarive, Académie Malgache.
Chapus, G.S. 1925. *Quatre-vingts années d'influences européennes en Imerina (1815–1895).* Tananarive: Académie Malgache.
Chapus, G.S. and Birkeli, E. 1944–45. "Historique d'Antsirabe jusqu'en l'année 1905." *Bulletin de l'Académie Malgache* 26: 59–82.
Chapus, G.S. and Dandouau, B. 1951–52. "Les anciennes industries malgaches." *Bulletin de l'Académie Malgache* 30: 45–70.
Chardenoux. 1816. "Journal du voyage fait dans l'intérieure." British Library. Add.18129.
Chauvin, J. 1939. *Jean Laborde 1805–1878.* Tananarive, Imprimerie moderne de l'Émyrne, Pitot de la Beaujardière.
Chauvin, J. 1959. *Jean Laborde 1805–1878.* Tananarive: Mémoires de l'Académie Malgache.
Copland, S. 1822. *A History of the Island of Madagascar.* London: Burton and Smith.
Coppalle, A. 1825–6. "Voyage dans l'intérieur de Madagascar et à la capitale du roi Radame pendant les années 1825 et 1826." http://www.bextes.org/coppalle.pdf (accessed 15/10/07).
Couland, D. 1973a. *Les Zafimaniry, un groupe ethnique de Madagascar à la poursuite de la forêt.* Tananarive: Imprimerie Fanontam-Boky-Malagasy.
Couland, D. 1973b. "Les forgerons à la lisière forestière du Nord-Betsiléo (Madagascar)" *Journal de la Société des Africanistes* 43. 2: 235–242.
Decary, R. 1947–48. "La population de Madagascar." *Bulletin de l'Académie Malgache* 28: 29–47.
Decary, R. 1966. Coutumes guerrieres et organisation militaire chez les anciens malgaches. Paris: Editions Maritimes et d'Outre-Mer. 2 vols.
Descartes, M. 1846. *Histoire et géographie de Madagascar depuis la découverte de l'île, en 1606, jusqu'au récit des derniers événements de Tamatave.* Paris: P. Bertrand.
Dez, J. 1970. "Éléments pour une étude de l'économie agro-sylvo-pastorale de l'Imerina ancienne" *Université de Madagascar: École nationale supérieure agronomique* 8: 9–160.
Dumaine. 1790. "Voyage au pays d'ancaye, autrement dit des Bezounzouns, Isle de Madagascar." British Library. Add. 18128.
Ellis, W. 1838. History of Madagascar. Second volume. London: Fisher, Son & Co.
Ellis, W. 1867. *Madagascar Revisited.* London: John Murray.
Ellis, W. 1870. *The Martyr Church: A Narrative of the Introduction, Progress, and Triumph of Christianity in Madagascar; with notices of personal intercourse and travel in that island.* London: John Snow.
Ellis, W. 1859. *Three Visits to Madagascar during the years 1853–1854–1856. Including a Journey to the Capital. With Notices of the Natural History of the Country and of the Present Civilisation of the People.* London: John Murray.
Ellis, W. and Freeman, J.J. 1842. *Madagascar and its Martyrs: A Book for the Young.* London: John Snow.
Fontoynont, A.M. and Nicol, H. 1940. *Les traitants français de la côte est de Madagascar – de Ranavalona I à Radama II.* Tananarive: Imprimerie moderne de l'Émyrne, Pitot de la Beaujardière
Freeman, J.J. 1830. "General Notice of Madagascar." Letter I. Concluded. 06/11/1830. *South African Commercial Advertiser* 6. 355.
Freeman, J.J. 1831. Freeman to Hankey, Tananarivo. 12/10/1831. School of Oriental and African Studies, Council of World Mission Archives. Madagascar Incoming Letters Bx.4 F.1 J.C.

Freeman, J.J. and Johns, D. 1840. *A Narrative of the Persecution of the Christians in Madagascar*. London: John Snow.
Gade, D.W. 1996. "Deforestation and Its Effects in Highland Madagascar." *Mountain Research and Development* 16. 2: 101–116.
Grandidier, A. 1916. "Souvenirs de voyages, 1865–1870." *Documents anciens sur Madagascar* 6.
Grandidier, A. and Guillaume. 1908. *Histoire Physique, Naturelle et Politique de Madagascar. Ethnographie de Madagascar*. vol.4. tomes 1–3. Paris: Imprimerie Nationale.
Grandidier, G. 1928. *Histoire Physique, Naturelle et Politique de Madagascar. Ethnographie de Madagascar*. vol.4. tome 4. Paris: Hachette et Société d'Éditions Géographiques, Maritimes et Coloniales).
Griffiths, D. 1843. *Hanes Madagascar*. Machynlleth: Richard Jones.
Hanson, P.W. 2008. "Governmentality, Language Ideology, and the Production of Needs in Malagasy Conservation and Development" *Cultural Anthropology* 22. 2: 244–284.
Hastie, J. 1817. "Diary." Public Record Office. CO.167/34.
Hastie, J. 1821. Hastie to Griffiths, Port Louis, 18 February 1821. School of Oriental and African Studies, Council of World Mission Archives. Madagascar Incoming Letters. Bx.1 F.2 J.B.
Hastie, J. 1822. "Diary." Public Record Office. CO.167/63.
Hastie, J. 1824–25. "Diary." Public Record Office. CO.167/78.
Hewlett, A.M. 1887. "Mantasoa and its Workshops; a page in the History of Industrial Progress in Madagascar" *Antananarivo Annual and Madagascar Magazine* 11 (1887): 295–300.
Hilsenberg, C.T. and Bojer, W. 1833. "A Sketch of the Province of Emerina, in the Island of Madagascar, and of the Huwa, its Inhabitants; written during a Year's Residence" (Tannanarivou, Emerina, May 1823). In Hooker, William Jackson. Ed. *Botanical Miscellany*. London: John Murray. vol.3: 246–275.
Hugon. 1808. "Aperçu de mon dernier voyage à ancova de l'an 1808." British Library. Add.18137.
J.L.K. [1849]. *Madagascar and its Martyrs. A Book for the Young*. Philadephia: Presbyterian Board of Publication.
Johns, D. 1826. Johns to LMS, Tananarive, 8 November 1826. School of Oriental and African Studies, Council of World Mission Archives. Madagascar Incoming Letters. Bx.2 F.3 J.C.
Johns, D. 1832. Johns to Arundel, Tananarivo. 15/04/32. School of Oriental and African Studies, Council of World Mission Archives. Madagascar Incoming Letters. Bx.4 F.2 J.C.
Jones, D. 1820a. "A Journal to Madagascar in 1820." Madagascar Journals. Madagascar and Mauritius 1 (1816–24), School of Oriental and African Studies, Council of World Mission Archives.
Jones, D. 1820b. Jones to Hastie. 23/10/1820. Quoted in Hastie to Griffiths, Port Louis, 18 February 1821. Archives of the National Library of Wales), Aberystwth. 19157E.
Jordaan, B. 1969. *Splintered Crucifix; early pioneers for Christendom on Madagascar and the Cape of Good Hope*. Cape Town: C. Struik.
Laverdant, D. 1844. *Colonisation de Madagascar*. Paris: La Société Maritime.
Le Sage. 1816. "Mission." Public Record Office. CO.167/34.
Leminier, N. 1970. "Notes sur une excursion faite dans l'intérieur de l'ile de Madagascar en 1825" *Bulletin de Madagascar* 292.
Lescalier. 1792. "Voyage à l'isle de Madagascar." British Library. Add.18128.
Lloyd, J.A. 1850. "Memoir on Madagascar." *Journal of the Royal Geographical Society of London* 20: 53–75.
Lyall, R. 1954. "Journal" (1827–8). Eds. G.S. Chapus & G. Mondain. Tananarive: Académie Malgache.
Mason, K. 1959. "Customs and Habits of the Merina Tribe of Madagascar." unpublished ms. Leicester.
Mayeur, N. 1775. "Voyage dans le nord de Madagascar." British Library. Add.18128.
Mayeur, N. 1777. "Voyage au pays d'ancove, autrement dit des hovas ou Amboilamba dans l'intérieur des terres, Isle de Madagascar." British Library. Add.18128.

Mayeur, N. 1785. "Voyage au pays d'ancove, par le pays d'ancaye autrement dit des Baizangouzangoux." British Library. Add.18128.

Michel, L. 1957. *Moeurs et coutumes des Bara*. Tananarive, Imprimerie officielle.

Mullens, J. 1875. *Twelve Months in Madagascar.* London; James Nisbet.

Nativel, D. 2005. *Maisons royales, demeures des grands à Madagascar*. Paris: Karthala.

Nilsen, J.A. [1943]. "Ny tantaran'ny distry Fianarantsoa, 1878–1943," 7 – ms. 57H, Archives of the Fiangonana Loterana Madagaskara/ Norwegian Missionary Society, Isoraka, Antananarivo, Madagascar.

Oliver, S.P. 1881. *On And Off Duty, Leaves From An Officer's Note-Book*. London: W.H. Allen.

Oliver, S.P. 1886. 2 vols. *Madagascar; an historical and descriptive account of the island and its former dependencies*. London : Macmillan.

Pfieffer, I. 1861. *The Last Travels of Ida Pfeiffer: Inclusive of a Visit to Madagascar: with a Biographical Memoir of the Author*. London: Routledge, Warne and Routledge.

Ralalarimanga, H. "De l'oralité à l'écrit : droit et gestion durable des ressources naturelles renouvelables de la forêt de Merikanjaka" *Taloha* 19 (2010), fn. http://www.taloha.info/document.php?id=789 (accessed 31/08/2011).

Raminadrasoa, F. and Ratsivalaka, G. 1979. "De l'usage du concept "ethnie" dans l'historiographie de Madagascar. L'exemple des Bezanozano au XVIIIe siècle" *Colloque d'histoire,* Université de Madagascar. Toliara.

Razafintsalama, A. 1983. "Les funérailles royales en Isandra d'après les sources du XIXe siècle." in Raison-Jourde, Françoise. Ed., *Les souverains de Madagascar.* Paris: Karthala.

RPPD – Readiness Preparation Proposals Development Team, Madagascar. 22/01/2010. http://www.forestcarbonpartnership.org/fcp/sites/forestcarbonpartnership.org/files/Documents/PDF/Mar2010/ENGLISH_RPP_Mada_Main_plus_annexes_v6_with_disclaimer.pdf (accessed 26/05/10).

Robequain, R. 1958. *Madagascar et les bases dispersées de l'union française*. Paris: PUF.

Raombana. 1853a. "Annales." Archives de l'Académie Malgache, Antananarivo.

Raombana. 1853b. "Histoires." Archives de l'Académie Malgache, Antananarivo.

Raombana. 1980. *Histoires* (1853). Ed. Simon Ayache. Fianarantsoa: Librairie Ambozontany.

Ryan, Vincent W. 1864. *Mauritius and Madagascar. Journal of An Eight Years' Residence in the Diocese of Mauritius, and of a Visit to Madagascar*. London: Sheeley, Jackon, and Halliday.

Sémerville. 1933. "Souvenirs de Sainte-Marie" (1824) in *Bulletin de l'Académie Malgache* 16: 17–56.

Sibree, J. 1915. *A Naturalist in Madagascar: a Record of Observation, Experiences, and Impressions made during a Period of over Fifty Years' intimate association with the Natives and Study of the Animal & Vegetable Life of the Island.* London: Seeley.

Sibree, J. 1923. *Fifty Years in Madagascar*. London: Allen & Unwin.

Tacchi, A. 1892. "King Andrianampoinimerina, and the Early History of Antananarivo and Ambohimanga." *Antananarivo Annual and Madagascar Magazine* 16: 474–496.

Tourte, R. 2003. "Explorateurs et marchands à la recherche De l'Eldorado Africain, 1800–1885/1890." FAO – ftp://ftp.fao.org/docrep/fao/009/a0217f/a0217f02.pdf (accessed 20/05/10).

Toy, R. 1875. "The Late Mr. James Cameron: His Life and Labours (a funeral address)" *Antananarivo Annual and Madagascar Magazine* 1: 48–55.

Vassière, P. de la. 1885. *Vingt ans à Madagascar*. Paris: Lecoffre.

Wills, J. 1885. "Native Products used in Malagasy Industries." *Antananarivo Annual and Madagascar Magazine* 9: 91–98.

Wright, H.T. ed. 2007. *Early State Formation in Central Madagascar.* Ann Arbor: University of Michigan Museum of Anthropology.

# SOLVING MADAGASCAR: SCIENCE, ILLUSTRATIONS, AND THE NORMALIZING OF FAUNA OF NINETEENTH CENTURY MADAGASCAR

Thomas Anderson

Madagascar, the fourth largest island in the world (it is often described as a mini-continent) lies only a few hundred miles off the eastern coast of southern Africa, but possesses a unique ecosystem: over 80 percent of the species are endemic to the island (Goodman and Benstead 2003). From the time that Vasco da Gama rounded the Cape of Good Hope in 1497, European voyagers commented on Madagascar's peculiar life forms, and by the nineteenth century, when the first major scientific expeditions in the island occurred, its reputation as a land of unusual species was firmly established. Madagascar's political history accentuated its environmental uniqueness as it remained free from overt European control until relatively late. While during the 1820s Protestant missionaries established a relationship with the Merina kingdom of the high central plateau, which forged diplomatic ties with both Great Britain and France, the Merina crown entered an isolationist phase from 1835 to 1861 that limited Western exploration of the island. The reopening of Madagascar to Westerners in the 1860s led to an influx of naturalists, particularly missionaries, but their activities and itineraries continued to be regulated by the Merina government.

Those who explored Madagascar in the nineteenth century entered a world of discovery, finding not just an occasional but many new species. Madagascar was a world unto itself. The few individuals permitted to travel and collect data and specimens developed an acute sense of the uniqueness of the island, reporting their findings to scientific centers in London or Paris, including museums and botanical gardens, and in personal and professional correspondences with naturalists elsewhere in the world. Indeed, Madagascar's unique ecosystem placed it at the centre of some of the most important scientific debates of the nineteenth century—from Darwinism and race theory to biogeography and human origins. The major issues concerned the fact that, while Madagascar was geographically close to Africa, and possessed some species related to those of the African continent, it also had some that were related to Asian fauna; that many characteristically African fauna were absent from the island; and that it possessed a plethora of species that appeared to have no relationship to species elsewhere in the world.

Two major schools of thought emerged about Madagascar's fauna: the first emphasized its uniqueness; while the second attempted to relate it to species elsewhere. This chapter traces the rise of both schools of thought, the nature of the debate between them, and the crucial role played in that debate by illustrations.

## *Madagascar: A Unique Natural Enclave*

While European naturalists exploring the world regularly sent back to Europe new species from the entire animal kingdom, it was mammals that most captured the attention of the scientific world and the general public. In this context, the island of Madagascar, lying off the east coast of Africa, proved a puzzle for, despite its proximity to the African continent, Madagascar lacked giraffes, elephants, lions, rhinos, zebras, and other characteristically 'African' mammals. Naturalists found unusual mammals that similarly could only be understood in their Malagasy context. One of these was the tenrec, a shrew like animal that fed on insects, and which hibernated not in winter but in summer, during the dry season—a habit that made sense only when Madagascar's climate was factored in to understand that this was when food supplies were lowest (Milne-Edwards and Grandidier 1872; Cassel 1871: 189–90). Another unusual mammal was the fossa, Madagascar's only large predator for which, when first obtained by the British Museum in 1864, a whole genus had to be created (Gray 1872: 206–7). Such endemic species demonstrated the very special nature of Madagascar's fauna.

Naturalists, however, were as interested in those species absent from Madagascar. The missionary and naturalist James Sibree (1836–1929) wrote, "the fauna of the country does…contain some most interesting and exceptional forms of life, but it is almost as remarkable for what is omitted in it as for what it contains" (Sibree 1880: 38–9). For a scientific world eager to establish linkages between species and regions, the absence of so many species that characterized the African continent, from which Madagascar was separated by only a few hundred kilometers of water, was deeply perplexing.

As Western naturalists in Madagascar collected and cataloged specimens, they attempted to reconcile their growing knowledge of the local environment, and their discoveries of the island's peculiar life forms, with existing scientific theories. In so-doing, they posed the basic questions:

did Madagascar form a link, either between Africa and Asia, or between an ancient and a contemporary world, or did it represent a world unto itself? As Richard Baron, an English missionary naturalist commented in 1894, on the eve of French colonization of the island, "the fauna of Madagascar is of the most interesting and remarkable character, no country in the world, indeed, affording more anomalous forms of organic life" (Baron 1890: 198–9).

Indeed, some naturalists began to view Madagascar not as a world where species had developed in isolation so much as a throwback, a primitive enclave where, due to some quirk of fate, species long since extinct elsewhere continued to exist (Baker 1882: 30–47). Thus F.E. Beddard stated in 1900: "That island [Madagascar] is infested with mammals which occur nowhere else, and mammals, moreover, that ought, so to speak, to have effaced themselves from creation long since" (Beddard 1900: 795–6). An attitude emerged from this line of thinking that Madagascar's fauna had little to add to the debate about natural selection, and were 'primitive' in every sense. Hence the tenrec was portrayed as stupid and inferior, an "unaggressive" creature that had survived through its nocturnal burrowing nature and lack of competition. Indeed, if it entered a "competition" with a cat, "the tenrec and its kindred generally would have been swept away long ago by sheer force of mind" (Beddard 1900: 795–6). Again, the Grandidiers considered that there was something "inferior" about Madagascar's animal life (Grandidier 1902: 158–62). The island's fauna had found refuge on the island instead of adapting to it. Advocates of this perspective, wherein Madagascar represented a biological dead-end, considered that it was pointless to study Malagasy animals in terms of their environment.

In this debate, lemurs—notably the aye-aye—came to play a central role. The most salient of Malagasy mammals, lemurs dominated the great tropical forests of Madagascar, and for most Western naturalists came to define the island's natural life. Lemurs provided tangible proof of a Malagasy link with Africa and Asia. However, naturalists were taken aback by their diversity for, of the forty-eight known species of lemur that existed worldwide, thirty-six were unique to Madagascar. As naturalist Philip Sclater remarked, it was the fact that lemurs constituted "nearly one half of the whole mammalian-fauna of Madagascar, that renders the Malagasy fauna so very different from that of any other part of the world's surface" (Sclater 1899: 236–8).[1] Western naturalists were puzzled at why

[1] The most complete work was done by French naturalists Alfred Grandidier and Alphonse Milne-Edwards—see Grandidier 1892.

Madagascar should possess so many lemur species, yet lack other mammals prominent on the African continent, such as apes or gorillas.

Moreover, whereas conventional theories about lemurs stressed their habits and diet to be strictly arboreal (Cassell 1871: 400–1), naturalists discovered that in Madagascar lemur species differed widely. For example, ring tailed lemurs lived not in trees but on treeless slopes and rocks where they moved about with such ease that it was clear that they had become "pre-eminently adapted for this kind of locomotion" (Shaw 1885: 247–8). In addition, some lemurs were nocturnal, others diurnal; some fed on berries and fruit, others ate insects, spiders, and beetles (Shaw 1885: 247–252; Idem 1882a: 48–50; Grandidier 1902; Cowan, 1881: 54; Sibree 1894: 70–83). For some naturalists, lemurs represented an ancient form of life that had managed to survive because of isolation, not out of a battle for survival. Thus Sclater wrote, "Though allied to the monkeys, they have none of their vivacity and intelligence" (Sclater 1899: 105).

In this debate, illustrations played a critical role. While the purported aim of illustrations was to represent a species with accuracy and anatomical precision, naturalists who argued for the uniqueness of Malagasy fauna used drawings that focused on their abnormal, eerie, even frightening characteristics. Carl Linnaeus (1707–78), the Swedish botanist, zoologist, and physician who is known as the father of modern taxonomy, chose the name 'lemur' from the Latin *lemures*, signifying ancestral ghosts, because of its reputation for unnerving people at night with its 'ghostly' face, while tales circulated of lemur cries that "pierced the silent forests" of Madagascar and haunted the night, often unsettling travelers (Baron 1882: 164–79; Standing 1887: 18; Shaw 1885: 247–52).

Illustrations of the aye-aye were central to the debate over the uniqueness of Madagascar's fauna. A nocturnal primate found nowhere else in the world, the aye-aye was difficult to find, let alone capture and study. Illustrations of the animal served as shorthand, as a way to disseminate a particular message about Madagascar. As the comment below, one early illustration of the aye-aye noted: "The above figure of this singular animal will, in part, save the necessity of verbal description" (*Family Magazine* 1840: 298). Its unusual anatomical features, including a long, skeletal middle finger, baffled the scientific community. As the Madagascar naturalist Richard Baron explained, the aye-aye's "structure is so anomalous that it is difficult to know what position to give it in any system of classification" (Baron 1890: 200).

The French naturalist and explorer Pierre Sonnerat (1748–1814) was the first European to study the animal in detail. He obtained a couple of aye-aye specimens in Madagascar in 1780 and, although they died within a few months, Sonnerat's description of them provided the basis for the bulk of Western information on the animal for the next eighty years because no new specimens came to hand until the mid-nineteenth century. It was Sonnerat who bestowed the name "aye-aye" on the animal. Possibly following Linnaeus he explained the name by claiming—erroneously—that upon seeing such a bizarre creature for the first time, the local Malagasy people exclaimed with surprise "aye-aye" (Sonnerat 1806: 137–9; see also Owen 1863a: 3–4).[2]

For a considerable time after Sonnerat's 1780 visit to Madagascar, no new specimens of the animal were either sighted or captured. In the absence of specimens, naturalists were divided as how to classify the aye-aye. Speculation about the aye-aye's unusual anatomy reflected a wider debate about how to conceptualize the animal and Madagascar. This in turn was often a question of interpretation, which inevitably reflected cultural values (Ritvo 1997). Some naturalists, such as Georges Buffon and Georges Cuvier, classified the aye-aye as a member of the rodent family, primarily on the basis of the teeth. Others classified it within the lemur family (Buffon and Cuvier 1826: 262–5; Cuvier and Griffith 1835: 196–8). Yet others considered that the aye-aye had become extinct; some claimed that it had never existed, that it was mythical. Such speculation underscored the West's limited knowledge of, and access to, Madagascar.

Without firsthand knowledge, images of the aye-aye only loosely corresponded to its reportedly unusual anatomy. The images used to illustrate descriptions of the aye-aye in natural histories reflected the wider debate over perceiving Madagascar as a land of the strange or of the familiar. In Sonnerat's illustrations, the aye-aye's skeletal finger and long tail are reproduced, but otherwise the image is largely a construct as it was drawn without reference to a real specimen. In the illustration, the creature has a snarling face, with a large yet almost featureless ear, and sits as a primate would rather than standing on all fours. The sparse background only exaggerates the animal's size, while its sloping shoulders and thick, raised arm provide a muscular and threatening tone. This reflected the perceptions of Sonnerat that the aye-aye and its island environment were alien, even frightening. Unsurprisingly, given the absence of reference specimens, these characteristics were often reproduced in other early illustrations of

2 Later naturalists, such as the British naturalist Richard Baron, would dispute this claim.

the aye-aye.[3] For those who saw the aye-aye as a rodent, its teeth were definitive and their illustrations often focused solely on the animal's dental structure: Thus one image shows the aye-aye with a pointed muzzle and two protruding teeth. By contrast, the images produced by those who claimed the aye-aye was more akin to the lemurs emphasized its living habits and elongated finger.

Figure 1. Pierre Sonnerat's aye-aye.

3 For example, the image appeared in works such as John Frost (Frost 1848: 22), as well as the often read William Ellis (Ellis 1859: 142).

### *Madagascar Related to the Wider World*

The debate, and the illustrations, changed dramatically from 1859 when the English comparative anatomist and paleontologist *Richard Owen* (1804–92) received a fresh aye-aye specimen from Humphrey Sandwith (1822–81), the British colonial secretary on Mauritius. Sandwith, who had obtained it live from a local source on Madagascar, first made notes about its behavior—the first such observations since Sonnerat in 1780—then killed it with chloroform, pickled it in a jar of alcohol, and shipped it to Owen in London (Owen 1863a: 7–8).[4] In conducting what constituted the first anatomical study of the creature, Owen noted that the aye-aye was three and half feet in length and so resembled in size a "domestic cat." Further, by carefully comparing the aye-aye with the rodent and lemur families, and paying particular attention to the specimen's limbs, Owen found that its "nearest resemblance" was to the lemur: He consequently classified them as distant relatives (Owen 1863a: 10, 46).

However, Owen dismissed suggestions that the aye-aye offered evidence for the theories of other prominent naturalists, including the notion of natural selection advanced by Charles Darwin (1809–82) (Rupke 1994: 242–8). Indeed, he argued that it strained credulity to believe that bird-eating lemurs could adapt over time to change into a wood-boring, larvae eating animal (Owen 1863b: 114–6). Rather, Owen argued that the aye-aye offered a prime example of his own theory regarding species development and the "ordained potentiality of second causes." He considered that there was a "design" behind the development of the aye-aye, but confessed "ignorance" over the process involved (Owen 1863a: 65).

This context explains why, although Owen's study provided a template for the visual representation of the animal, illustrations of the aye-aye (and other examples of Malagasy fauna) by other naturalists shifted according to whether the author wished to present Madagascar as a familiar or a strange land. In his, *Monograph on the Aye-aye*, Owen used the renowned German illustrator of natural history, Joseph Wolf (1820–99), whose pictures of the aye-aye were subsequently widely copied and adapted. Wolf viewed himself as a naturalist and artist priding himself on producing scientifically exact yet artistic representations that represented animals in their natural setting. This was no mean feat for two

[4] Sandwith's reports were confirmed by the French naturalist Auguste Vinson who had traveled to the island in 1862 and studied an aye-aye he imported to Reunion – Vinson 1865: 556–7.

main reasons. First, the aye-aye's nocturnal habits made it difficult to precisely observe its movements. Wolf wrote of nocturnal animals that "you must know the structure of the animal, and then you must learn to make it do what you like...you must become an Impressionist." Second, as "one of the greatest rarities," there were few live specimens of the aye-aye, so that Wolf often had to manipulate dead specimens into "live" poses for his sketches (Palmer 1895: 102–4, 187–9). He focused on the anatomy, and prominently displayed the elongated finger which he showed searching for food. He also produced highly detailed drawings of its hands next to a full life-like image of the aye-aye. In other illustrations portraying the aye-aye standing on a tree limb, its anatomical features are also clearly delineated. In these, Wolf concentrates on the aye-aye's large ears which he pushes forward; its eyes, accentuated by the small pupil surrounded by a large orb of white; its skeletal fingers, which are flexed as if in motion; and its snubbed nose. All of these attributes enhance the viewer's sense of aye-aye's strangeness (Owen 1863a).

However, Wolf sometimes represented the aye-aye as a more familiar, even sympathetic creature. In one drawing of an aye-aye drinking from a pool of water, Wolf recreated a scene described to him by Sandwith. Here the aye-aye possesses a sad facial expression, with drooping ears and

Figure 2. Joseph Wolf's aye-aye.

lowered eyes as it cautiously drinks. Its limbs are withdrawn into its body, and its tail hidden from view, suggesting a demure creature. Finally, in 1862, just before it died, Wolf studied a live aye-aye at Regent's Park Zoo in London (Palmer 1895: 102–3). It seems that Wolf used candlelight to draw a close-up of the aye-aye's head and hands. This illustration comprises a

Figure 3. Joseph Wolf's aye-aye drinking.

Figure 4. Joseph Wolf's facial profile of aye-aye.

juxtaposition between the creature's unusual skeletal hands, and a softer, more familiar depiction of its head, with a serene facial expression, and almost catlike ears, eyes, mouth, and nose. Here, the aye-aye was transformed into a far less menacing and more familiar creature (Owen 1863a; Rupke 1994: 242).

Wolf's images were widely reproduced or copied in other publications and exhibitions, both those where the emphasis was on establishing the aye-aye as unique, as in the book on Madagascar written by French naturalist Auguste Vinson, Scribner's Monthly, a popular "Illustrated Magazine for the People" published in America (1870–81),[5] and those representing the aye-aye as a more familiar creature (Owen 1863a. plate 6; Rupke 1994: 244).

Owen and Wolf nevertheless created the environment for a lively academic debate in the late nineteenth century between naturalists who continued to argue for the distinctiveness of their individual discoveries, and those who subscribed to global scientific theories.[6] The use of images to craft a message about Madagascar was at the center of Western understanding of the island's natural history. The precision and realism of natural history illustrations provided them with scientific authority and a sense of objectivity; they were a means of advancing knowledge (Donald 2007; Blum 1993; Knight 1977). Illustrations were often the composite of many specimens—creating an aggregate average representation—rather than the image of a specific individual (Fa-ti Fan 2004: 49–52).

In the 1860s and 1870s, naturalists who maintained that Malagasy flora and fauna were exceptional advanced the theory that Madagascar and surrounding islands were remnants of a lost continent, one that in 1864 English zoologist and museum director Philip Sclater (1829–1913) termed "Lemuria" (Sclater 1864: 212–9).[7] Advocates of this school of thought envisioned Madagascar as a both unique and self contained enclave which could not be understood through comparison with the world outside. Accompanying Sclater's article was an illustration designed to impress upon readers the peculiar qualities of Madagascar. An aye-aye takes center stage, its large eyes and ears prominently displayed, and its long skeletal fingers grasping a tree trunk. Above the aye-aye, four lemurs peer out at the viewer while a tenrec crouches in the foreground (Sclater 1864: 212–9).

[5] The image is from Wolf, published in Owen 1863. plate 2, and was republished widely including, Vinson 1865. plate 1; *Scribner's Monthly.* 1871: 33; Mivart 1871: 122.

[6] For more on the wider debate between naturalists in the field and those in the metropole see, Endersby 2008.

[7] The French naturalist and Madagascar expert Milne Edwards advocated that the neighboring Mascarene islands represented a lost continent, although the absence of mammals suggested that Madagascar had been separate from it—Milne-Edwards 1873; see also *Harper's New Monthly.* 1872: 627. For a fuller look at the issue of Lemuria see Ramaswamy 2004.

Figure 5. Sclater's Malagasy fauna.

Again, Sibree, who spent much of his long life extolling the unusual qualities of Madagascar, in *The Great African Island* described and illustrated its "unique" flora and fauna. These illustrations included the aye-aye that, unsurprisingly, is portrayed as a wild, predatory creature, presented without a background, and its features drawn with sharp angles to enhance their unusual features. It possesses a fierce, even menacing face with large eyes, upright and taut ears upright, and fur composed of sharp, spiky individual hairs. Moreover, it is crouched low, with its fingers outstretched and flexed, and, with a curved spine, rigid tail, and legs arched, ready for action (Sibree 1880: 42–3). Sibree argued that the creature offered confirmation of the theory of natural selection—that "The whole organization of the aye-aye presents one of the most striking examples known of animal structure modified to serve special ends" (Sibree 1894: 503).

Most naturalists, then, disagreed with Owen. Darwin observed that Owen's analysis of the aye-aye offered "nothing new" (Darwin Project 2011), while Darwin's supporters considered that Owen's attempts to debunk natural selection "fell flat" (Huxley 2011).[8] Indeed, the majority

AYE-AYE (Cheiromys Madagascariensis)

Figure 6. James Sibree's aye-aye.

[8] Indeed, Owen's paper on the aye-aye aroused a small flurry of letters between Darwin and his friends, with Darwin asking to see the paper and friends including T.H Huxley, JD Hooker, JS Gray, and Charles Lyell, all prominent men of science, dismissing it. Both Grey and Huxley wrote to Darwin about Owen's paper, ridiculing Owen's claims but angry over his attempts to present an ulterior theory of selection, see Gray 2011; Huxley 2011.

view was that the aye-aye offered a classical example of a species adapting to local conditions. Indeed, rather than focus on the unique, most naturalists sought to find similarities between species, to incorporate local findings into a larger framework and thus to construct connections between Madagascar and the wider world that helped establish a sense of scientific progress (see Browne 1983; MacLeod 2001; Numbers and Stenhouse 1999). This majority view largely subscribed to the emerging biogeography school of thought that, drawing from geology, zoology, and botany, argued that all species developed from closely allied species which could thus be categorized into geographically distinct zones (Browne 1983: 224–5; Camerini 1993). Naturalists of the time engaged in creating global theories often rendered Madagascar's fauna in a more universal light that played down their uniqueness, minimized strange and unusual features, and depicted more familiar looking animals.[9] Their illustrations tamed fauna, often removing the animal from its environment.[10]

For example, Welsh-born Alfred Russel Wallace (1823–1913) had at first embraced the unique status of Madagascar. He had even tentatively supported the idea of Lemuria. However, he subsequently embraced the concept of biogeography following which he carefully divided up the regions of the world according to types of species in order to explain the distribution of the world's fauna. He incorporated Madagascar into this system by linking it to the African continent. Specifically, he categorized the island as a sub region of his Ethiopian region—asserting that there was "no doubt that [sic] Madagascar being more nearly related to the Ethiopian than to any other region; but its peculiarities are so great, that, were it not for its small size and the limited extent of its fauna, its claim to rank as a separate region might not seem unreasonable" (Wallace 1876: 74). While describing the aye-aye as "one of the most extraordinary of the mammalia now inhabiting the globe," Wallace stated that it was a clear example of "an exceedingly specialized form" of species adaptation to its environment (Wallace 1876: 177).

Wallace's theories were reflected in the illustrations that accompanied his descriptions of fauna. For instance, in his *Geographical Distribution of Animals,* Wallace included a sketch labeled "Scene in Madagascar, with Characteristic Animals" in which lemurs are absent. He explained this by claiming that lemurs were "relatively well-known" (Wallace 1876: 278,

---

[9] For more on how taming animals became a moral duty see, Donald 2007. For an analysis on how imperialism reshaped the natural world to fit its needs see Drayton 2000.

[10] This was a common practice until the mid-nineteenth century, see Knight 1977; Dance 1978.

Figure 7. Alfred Russel Wallace's "Scene in Madagascar".

plate 6). Instead Wallace focused on lesser-known species, the peculiarities of which he explained to the reader by reference to general theories of the inter-relatedness of fauna. The aye-aye took pride of place in the forefront of the picture. However, its long skeletal finger, its most characteristic and anomalous feature, is less pronounced and hidden in shadow. Wallace noted that while "nocturnal and extraordinary" it probably represented "very ancient mammalian fauna." In the distance, Wallace placed river-hogs that were "allied to African species"—an indication of "a later immigration from the mainland than in the case of most of the other Mammalia" (Wallace 1876: 278). Again the birds he illustrated appear similar to species elsewhere. Throughout, Wallace's images of Madagascar's fauna serve to illustrate his theory that isolation led species to develop in different ways, but that all fauna were globally connected, those of Madagascar and the African continent being related.

In the latter part of the nineteenth century, greater Western exploration of Madagascar, and the arrival in Europe of new specimens led to greater familiarity with anatomy and activities of the aye-aye. French naturalists Alfred Grandidier (1836–1921) and Alphonse Milne-Edwards (1835–1900), and British naturalists Richard Baron (1847–1907) and George Shaw (1842–1917) added considerably to knowledge about, the aye-aye, including its dietary habits (the use of its long finger to extract grubs and larvae), its building of nests to rest and bear young, its powerful chewing ability, its elusiveness, and its close kinship with the lemurs (Humphrey Sandwith in Owen 1863a: 7–8; Bartlett 1863; Tegetmeier 1862; Grandidier 1890 and 1892; Pollen and Van Dam 1868, vol.2: 1–19; Vinson 1865: 558–61; Shaw 1882b; Shaw 1885: 247–52; Baron 1890: 196–211).

Most naturalists followed Wallace in considering that the aye-aye offered confirmation of the theory of natural selection. For them, the true problem arose not from analysis of a single species, but in attempting to find a general explanation for the unique nature of Madagascar's fauna and flora. Unlike Wallace, most chose not the aye-aye but the lemur to illustrate their theory. While Madagascar's abundance of lemur varieties helped establish it as a highly unusual island, naturalists subscribing to the majority view often characterized the lemur as possessing human qualities, with a child-like cry. Their illustrations of the lemur often enhanced the human features of its face in order to create a more familiar and sympathetic animal. Such anthropomorphizing was a popular technique in natural history illustrations, specifically designed to create a connection between human viewers and new or unusual species (Dance 1978; Donald 2007). Indeed, Grandidier and Milne-Edwards, who documented

Figure 8. Alfred Grandidier's lemur.

Figure 9. *Science's* aye-aye.

much of the early natural history of the island, and produced many of the first illustrations of lemur species, often cast the animals with smiling faces, and eyes gazing directly out at the onlooker, so as to promote a sense of familiarity for the human viewer. Such positions created a stylized setting. In one typical illustration, a lemur is sitting gazing at the viewer, while the thick bridge of its nose, close cropped fur on its head, and fingers with opposable thumbs look distinctly human. Similarly, the image of a lemur relaxing in a tree, smiling slightly outwards, dispels for the viewer any feelings of fear and strangeness (Grandidier 1892, vol. 20, plate 6).

Similarly, the illustration of the aye-aye in the influential journal *Science* (December 1881) that was subsequently often reproduced, the creature's anatomical features are retained, but it is portrayed as a friendlier, almost feline character. It has a slight smile, giving its face an almost kind expression. Its fur is softer, less individualized, and its ears slack and calm. Its relaxed limbs and body, and its drooping, almost flaccid tail, gives it a subdued, non-threatening appearance, one accentuated by its background—an idyllic environment with leaves and tree branches that creates a soothing atmosphere, and ensures that the aye-aye does not overwhelm the overall illustration (*Science* 1881).

In this article, as in many general scientific publications, there is an attempt to demonstrate not the uniqueness of the aye-aye, but rather its connection to a global whole. Such illustrations attempted to elicit sympathy from the viewer by focusing not on Madagascar or the lemur's unique abilities so much as its close relationship to other primates, and therefore humanity.[11] Although endemic to Madagascar and novel to the nineteenth-century reader, the lemur thus became familiar to the educated European.

## *Conclusion*

From the end of the eighteenth century Western naturalists were perplexed by Madagascar's abundance of strange fauna, and their inability to relate those animals to species found elsewhere in the world. The island had few of the species dominant on the neighboring African continent, but possessed creatures such as the aye-aye and lemur, that not only

[11] Dance, *Art of Natural History;* Donald, *Picturing Animals in Britain.*

possessed unusual anatomical features, but were unique to Madagascar. For such naturalists, finding a theory that could accommodate Madagascar's fauna was a priority. Two Schools of Thought emerged, one that emphasized Madagascar's fauna as an example of the existence of a unique number of primitive antediluvian species; and the other that argued for global theories that linked the island's unusual fauna with species elsewhere, notably on the neighboring African continent. Illustrations played a central role in the arguments of both sides. Thus the first school portrayed Malagasy fauna as strange, eerie, even frightening creatures, and emphasized to a maximum their unusual anatomical features and habits. They made little or no attempt to relate them to their environment. By contrast, the second global theory school, portrayed Malagasy fauna in a softer light, downplaying their unusual characteristics, and giving them features or expressions that made them look more familiar—such as eyes that looked directly outwards, engaging the onlooker, and slight smiles. Members of this school also placed Malagasy fauna in context, explaining their features and habits in terms of their environment. They argued that Madagascar's fauna represented primitive versions of African species, and thus categorized them as part of the African zone of biogeography.

By the late nineteenth century, the global theorists had emerged as the dominant school of thought. For example, Wallace argued that although Madagascar possessed a large number of unusual species, it was "better and more natural to keep it as a sub-region" of Africa (Wallace 1876: 75). Despite caveats in his narrative (see Grandidier 1902: 157–216), Wallace and the majority of his naturalist colleagues placed Madagascar within the African region of biogeography because otherwise it defied categorization (Wallace 1876: 74–5). They had made a deliberate choice, one that prized conformity and privileged the global over the local. Their central aim was to show Madagascar's fauna as explicable—and thus the island as a land capable of domestication. Domestication provided control. Species that defied explanation were beyond science and Western understanding. However, by defusing the strangeness of the fauna of Madagascar through kindly images, and by and stressing that Malagasy fauna had global connections, naturalists created a space that became understandable for Europeans, transforming Madagascar from a strange and dangerous enclave to one that could be mastered by science. The globalization of the fauna of Madagascar allowed Westerners to construct a narrative that portrayed Madagascar as a known and familiar land, laying the groundwork for scientific knowledge to later be used as part of the basis for colonial control.

## *Bibliography*


Baker, J.G. 1882. "On the Natural History of Madagascar." *Antananarivo Annual and Madagascar Magazine* 6: 30–47.

Baron, Richard. 1890. "A Malagasy Forest." *Antananarivo Annual and Madagascar Magazine* 14: 196–211.

Baron, Richard. 1882. "From Ambatondrazaka to Fenoarivo" *Antananarivo Annual and Madagascar Magazine.* 6: 164–79.

Bartlett, A.D. 1863. "Observations of the Habits of the Aye-Aye Living in the Gardens of the Zoological Society, Regent's Park, London." *Report of the Thirty Second Meeting of them British Association for the Advancement of Science*: 103–4.

Beddard, F.E. 1900. "A Rarity at the Zoo." *Chamber's Journal.* 34. 154: 795–6.

Blum, Ann Shelby. 1993. *Picturing Nature: American Nineteenth-Century Zoological Illustration.* Princeton, N.J: Princeton University Press.

Browne, Janet. 1983. *The Secular Ark: Studies in the History of Biogeography.* New Haven: Yale University Press.

Buffon, Georges Louis Leclerc and Cuvier, Georges. 1826. *Oeuvres complètes de Buffon.* Paris: Chez audouin frères etchez N. Delangle.

Camerini, Jane R. 1993. "Evolution, Biogeography, and Maps: An Early History of Wallace's Line." *Isis* 84. 4: 700–27.

Cassell. 1871. *Cassell's Popular Natural History* (London: Cassell, Petter, & Galpin).

Cowan, William Deans. 1881. *The Bara Land: A Description of the Country and People.* Antananarivo: L. M. S. Press.

Cuvier, Georges and Griffith, Edward. 1835. *The Animal Kingdom.* London: Whittaker and Co.

Dance, S. Peter. 1978. *The Art of Natural History: Animal Illustrators and Their Work.* Woodstock, N.Y: Overlook Press.

Darwin, Charles. 2011. Charles Darwin to JD Hooker, 17 March 1863, Letter 4048 – www.darwinproject.ac.uk (accessed 01/09/11).

Donald, Diana. 2007. *Picturing Animals in Britain, 1750–1850.* New Haven: Yale University Press.

Drayton, Richard. 2000. *Nature's Government: Science, Imperial Britain, and the 'Improvement' of the World.* New Haven: Yale University Press.

Ellis, William. 1859. *Three Visits to Madagascar During the Years 1853–1854–1856.* New York: Harper.

Endersby, Jim. 2008. *Imperial Nature: Joseph Hooker and the Practices of Victorian Science.* Chicago: University of Chicago Press. *Family Magazine* [*The*]. 1840.

Fa-ti Fan. 2004. *British Naturalists in Qing China.* Cambridge, MA: Harvard University Press.

Frost, John. 1848. *The Book of Travels in Africa.* New York: Appleton.

Goodman, Steven and Benstead, Jonathan. Eds. 2003. *The Natural History of Madagascar.* Chicago: University of Chicago Press.

Grandidier, Alfred. 1892. *Histoire Physique, Naturelle Et Politique De Madagascar.* Paris: Imprimerie Nationale.

Grandidier, Alfred. 1890. *Histoire physique, naturelle et politique de Madagascar. 10, Histoire naturelle des mammifères* T. 5. Paris: Hachette.

Grandidier, Guillaume. 1902. "Zoologie" In Raphael Anatole et al. Eds., *Madagascar au debut du xxe Siecle.* Paris: F.R. de Rudecal et cie: 158–62.

Gray, J.E. 1872. "Notes on a New *Propithecus* and the Fossane from Madagascar" *The Annals and Magazine of Natural History* 4. 57: 206–7.

Gray, J.S.ss 2011. J.S. Gray to Darwin, 29 Jan 1862, Letter 3419 – www.darwinproject.ac.uk (accessed 01/09/11).

*Harper's New Monthly.* 1872: 627.

Huxley, Thomas. 2011. T.H. Huxley to Charles Darwin 9 Oct 1862, Letter 3755 – www.darwinproject.ac.uk (accessed 01/09/2011).

Knight, David M. 1977. *Zoological Illustration: An Essay Towards a History of Printed Zoological Pictures*. Folkstone, Eng: Dawson.
MacLeod, Roy M. Ed. 2001. *Nature and Empire: Science and the Colonial Enterprise*. Chicago: University of Chicago Press.
Milne-Edwards, Alphonse. 1873. *Recherches sur la faune ornithologique éteinte des iles Mascareignes et de Madagascar.* Paris: G. Masson.
Milne-Edwards, Alphonse and Grandidier, Alfred. 1872. "Description d'un Nouveau Mammifère Insectivore de Madagascar (Geogale aurita)" *Annales* de Sciences Naturelle Zoologie. 15, art. 19: 1–5.
Mivart, St. George Jackson. 1871. *On the Genesis of Species*. London: D. Appleton.
Numbers, Ronald L. and Stenhouse, John. Eds., *Disseminating Darwinism: The Role of Place, Race, Religion, and Gender.* Cambridge, U.K: Cambridge University Press.
Owen, Richard. 1863a. *Monograph on the Aye-aye*. London: Taylor and Francis.
Owen, Richard. 1863b. "On the Characters of the Aye-Aye, as a Test of the Lamarckian and Darwinian Hypothesis of the Transmutation and Origin of Species." *Report of the Thirty Second Meeting of the British Association for the Advancement of Science*: 114–116.
Palmer, Alfred Herbert. 1895. *The life of Joseph Wolf, animal painter*. London: Longmans.
Pollen, Francois P.L. and Van Dam, D.C. 1868. *Recherches Sur La Faune De Madagascar Et De Ses Dépendances*. Vol. 2. *Mammiferes et Oiseaux*. Leyde, J.K. Steemhoff.
Ramaswamy, Sumathi. 2004. *The Lost Land of Lemuria: Fabulous Geographies, Catastrophic Histories*. Berkeley: University of California Press.
Ritvo, Harriet. 1997. *The Platypus and the Mermaid, and Other Figments of the Classifying Imagination*. Cambridge, Mass: Harvard University Press.
Rupke, Nicolaas. 1994. *Richard Owen: Victorian Naturalist*. New Haven, CT: Yale University Press. *Science* 1881. Vol. 2. No. 75: 574–6.
Sclater, William Lutley and Sclater, Philip Lutley. 1899. *The Geography of Mammals*. London: K. Paul, Trench, Trübner & Co.
Sclater, Philip Lutley. 1864. "The Mammals of Madagascar." *Quarterly Journal of Science* 1. 2: 212–9.
*Scribner's Monthly.* 1871. 2. 1: 33.
Shaw, George A. 1885. *Madagascar and France*. London: Religious Tract Society.
Shaw, George A.1882a. "A Few Notes Upon Four Species of Lemur." *Antananarivo Annual and Madagascar Magazine*. 6: 48–50.
Shaw, George A. 1882b. "The Aye-Aye" *Antananarivo Annual and Madagascar Magazine* 6: 123–124.
Sibree, James. 1894. "The Mammals of Madagascar: The True Lemurs." *Antananarivo Annual and Madagascar Magazine*. 18: 70–83.
Sibree, James. 1880. *The Great African Island*. London: Trubner & Co.
Sonnerat, Pierre. 1806. *Voyage aux Indes Orientales et a la Chine*. Paris: DENTU.
Standing, Herbert F. 1887. *The Children of Madagascar.* London: Religious Tract Society.
Tegetmeier, W.B. 1862. "The Aye-Aye." *The Intellectual Observer* 1: 379–80.
Vinson, Auguste. 1865. *Voyage à Madagascar*. Paris: Librairie Encyclopedique de Roret.
Wallace, Alfred Russel. 1876. *The Geographical Distribution of Animals*. New York: Harper & Brothers.

# *LEX LOCI* MEETS *LEX FORI*: MERGING CUSTOMARY LAW AND NATIONAL LAND LEGISLATION IN MADAGASCAR

Sandra J.T.M. Evers[1]

In 2004, the Malagasy national government of former President Ravalomanana introduced a new land registry system under the name *Programme National Foncier* (PNF), with the stated mission of improving the quality of agricultural practices by offering tenure security. With a growing population of over 21 million (World Bank, Data and Statistics, 2012), land is increasingly scarce. Next to its scarcity, land is facing rapid degradation due to erosion and over-use, which incites people to resort to long-standing *tavy* ('shifting cultivation') practices. These, however, are prohibited and subject to corrective penalties (Kull 2004). The anti-*tavy* legislation is closely associated with the former Ravalomanana government plan to improve the Malagasy economy by implementing a land registration system, with a view to increasing taxation revenues for the cash-strapped state treasury. The strategy was to offer Malagasy tenure security over their existing plots, the underlying aim being to ban expansion of agricultural fields through *tavy*.

The government expected land registration to act as an incentive to encourage farmers to invest in the upkeep of lands. Improved land quality would favour intensification of agricultural output while increasing state revenues through World Bank-inspired stratagems designed to encourage self-financing of state expenditures. The World Bank relies on the theories of Peruvian economist Hernando de Soto (2000) who argued in favour of capitalization of developing economies in order to integrate immovable property into legal systems to assess monetary value of such assets: "Because the rights to these possessions are not adequately documented, these assets cannot readily be turned into capital, cannot be traded outside the narrow local circles where people know and trust each other, cannot be used as collateral for a loan, and cannot be used as a share

[1] I am most grateful for the assessments of this chapter by the anonymous reviewers and Karen Middleton.

against an investment" (ibid.: 12). Land title, De Soto claimed, would confer tenure security, encourage further investment in land, and intensify production processes, which in turn would facilitate farmers' ability to honour their duty to pay taxes over the registered land (cf. Otto 2009: 176).

In practice, however, the land registry system has yet to be fully implemented.[2] Nor has land registration necessarily triggered increased investment by farmers in their lands. This chapter discusses the Malagasy legislation in terms of land registration and certification and attempts to identify some areas of possible future conflict due to the ongoing cleavages between positive law land legislation (*lex fori*) and land ownership based on custom (*lex loci*).[3] It should be stressed right from the outset that the *lex fori* – *lex loci* distinction is analytical here, as in the day to day people draw discursively and practically from varied social and legal frameworks. Kasanga and Woodman therefore refer to such practices as "living laws" because "The living laws of today are derived historically from the indigenously generated living laws which were followed before the inception of the colonial period, although the people have adapted these laws radically to modern circumstances" (2004:161). Also Malagasy legislation contains elements of "living laws" as will be argued below, however, it is important to note that codification of lived practices into national legislation makes such arrangements not only less fluid and adaptable to local practices, but also implies that if a remedy to perceived wrongs is sought, legal conflict resolution falls within the jurisdiction of the *lex fori* courts (see *infra*).

### *Programme National Foncier (PNF): The Quest to Harmonize Competing Jurisdictions*

In 2004, with the support of the World Bank, the Malagasy government launched the *Programme National Foncier*, PNF (Teyssier 2004: 1–3), which is still in force and effect today. PNF envisaged a decentralised registration system, where people could obtain their land certificates in

[2] In our 2007 publication (Pronk & Evers), we pointed to the conflicts arising out of legal and institutional pluralism. So far PNF has not delivered on the hope that its implementation could mitigate such conflicts (cf. Middleton chapter).

[3] *Lex fori* refers to the law of the forum, or court; that is, the positive law of the state, country, or jurisdiction of the court where a dispute is being dealt with. *Lex fori* controls all matters pertaining to remedial or procedural rights, as opposed to substantive rights. *Lex loci* refers to the law of the locality. The *lex loci* or 'law of the place' where rights are acquired or liabilities are incurred governs the substantive rights of parties to a dispute.

Local Land Offices (LLO or in French: *guichet foncier*). The Malagasy were formally informed of the new policy goals in the *Lettre de Politique Foncière*, which was ratified by government decree on 3 May, 2005. Four strategies of property reform were set out in this letter: 1) Restructuring, modernisation and computerising of the land registry and topographical database. This area consists of electronic date entries of land registry archives and investment to upgrade land registry equipment and regularise land titles; 2) Improvement and decentralisation of land management. The second area creates a local land property administrative infrastructure, the communal and inter-communal LLO authorised to deliver and process the *certificats fonciers* based on local, public and adversarial procedures; 3) Update of property and domain legislation, to comply with the new property system and to integrate old legal forms of property rights which do not reflect the status of the current land holders; 4) National training programme in property management and communications. This involves the training of local personnel in land titling and security, professional and university training. These four areas are complemented by the *Observatoire du Foncier*, a monitoring and review tool, which oversees actions of the PNF and the impact of measures under the new land tenure policies.

Prior to 2005, all land that was untitled was automatically deemed to be State-owned. The only way to legally secure land was to obtain a title deed delivered by the State land registry services. Under the new land legislation, two major changes have been implemented. Firstly, untitled but occupied land is no longer automatically property of the State. Land claimed by local people may be categorized as 'untitled private property'. This however in reality does not mean that these legal provisions prevent government claims to land under local use (Evers et al. 2011). Secondly, local government institutions (municipalities) have been granted new powers. By creating a local land registry office, it is responsible for recognizing private property rights and issuing land certificates. The applicant can be an individual but people can also certify their land as a group. Local claims to land are no longer assessed by the State administration but by a local commission[4] on the basis of pre-existing customary rights (Merlet & Bastiaensen 2012: 26).

[4] This local commission is comprised of elected representatives of the village and neighbors of the concerned claimant.

The PNF planned to convert provisional deeds of peasants (*petits papiers*)[5] into official certificates, which in turn could eventually become title deeds. The issuance of these documents would recognise local values in terms of property and the legal procedure within the framework of implementing a long-term plan intended not only to improve land security but also turn the land registry unit of the public administration into a more professional and functioning public service, which had been in steady decline for many decades. At the outset of the programme only about 10 % of land was registered, the existing national registration system was functioning poorly, titles were expensive and the processing period lengthy (Teyssier et al. 2010: 18).

Between 2006 and 2011, 416 Malagasy municipalities established land registry offices (Land Observatory 2011: 22), comprising approximately one-fifth of all municipalities. Since inception, land registry offices have received 120,000 certification requests and more than 60,000 certificates were delivered (ibid.) while the State land services only managed to issue on average 1,500 land titles per year. The average cost of a certificate is 15 USD and the process generally requires about six months. This stands in sharp contrast to the official title deed, which costs on average 507 USD and takes 6 years (ECR 2006, 2008; Teyssier et al., 2009).

It is important to note that the land titling procedure is still in existence, resulting in a dual legal land management system: the formalization of property rights through a title deed falls within the remit of the national government; the formalization of property rights to untitled land via certificates is a new power granted to local government (Teyssier 2010). In spite of these two options, most local land users have yet to secure their land claims through titles or certificates.

Thus, even with PNF, not all Malagasy are integrated into this positive law schema. Despite all the innovative measures, this presents one of the major challenges of the new system, as competing systems of ownership assessment and valuation constitute fertile ground for misunderstandings and disputes between the various stakeholders. These conflicts most commonly appear in three scenarios: 1) private versus private ownership disputes over land. These play out at the local level between families or within families. One likely case might concern an absentee (diaspora)

[5] At the local level people resorted to their own arrangements called *petits papiers*. These are written land transactions, signed by authorities like the mayor. However, they have little legal value and are solely acknowledged at the local level (Jacoby & Minten 2005: 6–7).

Malagasy asserting title to a specific property lot against generational land users or even an absentee landlord with no title but who claims to be the first to have cleared the land for cultivation notwithstanding use and cultivation of the same land by other families for generations; 2) national versus private conflicts originating in the local perception of land ownership and rights as compared to the positive law definition of it. The state may lay claims to land based on positive law legislation whereas local users might have no title or certificate but may nevertheless lay claim to the land based on the intergenerational use 3) these are indeed analytical divisions as a combination of elements mentioned under one and two is also common (data from Natural Resources research programme directed by Evers which commenced in 2005).[6]

### *The Legacy of Disputes Originating Prior to the PNF*

Before examining the effects of PNF on conflicts and particularly the avenues for resolving them, it is useful to examine potential conflicts rooted in earlier legislation, which continues to apply unless repealed. One statutory example provides us with an insight into the jurisdictional issue. The law states that land property rights are contingent upon ongoing exploitation of land. If rural land having a surface area larger than five hectares is not used for more than 5 years, the property reverts back to the state. This practice would appear to be the Malagasy version of the common law notion of escheat, and civil law of *prescription extinctive*. It should be noted that the five year period under the national law makes holding vacant land a highly precarious situation in rural areas.

> Art.3.- Where it has been established that a rural property with a surface area exceeding five hectares has not been occupied personally or at the cost of the owner or his successors or assigns, any such property shall be transferred and assigned to the State, who shall be deemed absolute owner thereof as of the date of such determination. (Law 74–021 of 20 June 1974).[7]

---

[6] In addition to the data compiled by the author, fieldwork information in this chapter stems from a joint research programme between the Department of Social and Cultural Anthropology (*VU University Amsterdam*) and the *Institut de Civilisations/Musée d'Art et Archéologie* (*Université d'Antananarivo*) financed by ICCO (Dutch Development NGO), which was initiated in January 2005. In this research, sixteen Dutch master students and sixteen Malagasy students jointly carried out fieldwork and research on natural resource management and poverty.

[7] Throughout the chapter translations of the legislation are by the author.

Following the statutory jurisdiction people can apply for title to land, which is not yet legally owned by somebody else, after a ten year period of cultivation of the land in question.[8] This might culminate in conflict at the local level between those who have vacated the land, but claim ownership to the land on the basis that their ancestors first cleared the land for cultivation (a classical category 1 conflict referred to above).

> Art.18. (*Ord. n. 62–047 du 20.09.62*) – With the exception of immatriculated lands or lands registered in the name of individuals or appropriated pursuant to duly constituted concession deeds or pursuant to public or private laws of general application, any occupant of Malagasy nationality who exercises personal, real, public and permanent control over land, evidenced either by buildings or by actual physical improvements thereupon, and such control is unopposed and uninterrupted, the whole in accordance with applicable custom and usage in the locality from time to time, for a period of ten years prior to the date of determination, may obtain a title deed to such property subject to the terms and conditions set forth in such deed, for property up to the maximum limit of 30 hectares (Law 60–004 of 15 February 1960).[9]

The pre-PNF statutory provisions provided for the acquisition of land title based on local practices; their legislative expression is derived from historical notions of adverse possession borrowed from Western legal systems, and how they recognise the acquisition of title through use and occupation. Thus, it should be noted that even under the pre-reform legislation, there existed considerable tension between generational land users and people who moved away from the areas while still laying claim to land based on first occupation of the territory.

---

[8] The general principle of adverse possession or acquisitive prescription for the benefit of persons having improved in an effective and permanent manner all or part of a property registered under the name of a third party was introduced in pre-reform legislation by Decree No 60–146 of 3 October 1960. But, as Razafarijaona (2007) points out in his legislative overview, occupants of lands during both the pre- and post-reform systems have rights which risk being defeated by superior title under positive law – either by way of a right of escheat and return of lands to the public domain, or by virtue of proper title held by a third party, which for all intents and purposes reduces the property rights of occupants to that of mere usufruct.

[9] The foregoing positive law notions correspond loosely to notions of adverse possession or escheat. See also article 3 of Decree N° 2007–1109 implementing law 2006–031, *fixant le régime juridique de la propriété foncière privée non titrée,* providing for recognition of a property right, "based upon individual or collective occupation which is real, public and permanent, in accordance with applicable custom and usage in the locality from time to time and the zoning of the land".

### *What is the Effect of PNF on Landowners' Rights?*

As stated above, PNF has the ambition of bringing all property rights within a positive law land scheme based on the registration and certification of land. A brief summary of some relevant provisions is in order, to be able to assess the consequences of PNF in terms of disputes and conflict resolution.

Law 2005–019 *fixant les principes régissant les statuts des terres* (hereinafter the "2005 Act") sets out the general PNF scheme applicable to all lands. PNF is based on the *Certificat foncier*, a form of land certificate, and the *Titre foncier*, or title deed. The 2005 Act definitions contain provisions that will prove central to the determination of rights in land disputes. The central concepts in this regard are the following:

The *Certificat foncier* "...recognizes a prima facie property right which is enforceable against third parties, subject to contrary proof." (Article 14 of the 2006 Act contains a similar provision.)

*Propriété* (*Droit de*) is the "real property right, which permits its user to use, enjoy and dispose of his property in his entire discretion, subject to existing legislation".

*Reconnaissance locale*: "legal procedure established at the local level...to determine and validate the legal status of land, by an ad hoc commission".

*Immatriculation*: "legal scheme applicable to real property rights which guarantees a final and unopposable title".

Article 23 of the 2005 Act states that immatriculation constitutes proof against third parties "of all real property rights" governing property. Article 27 in effect makes registration mandatory, stating that no real property right, lien, mortgage, charge or interest can be enforced against third parties "...unless it has been made public in the forms, conditions and limits set forth in this Act and its regulations". Article 29 then establishes the *titre foncier* as the supreme point of reference for all property rights in Madagascar, which is "final and unopposable; it constitutes the exclusive point of departure before Madagascar courts for the determination of real property rights, liens, charges and interests charging any property *to the exclusion of all other unregistered rights*" (emphasis added by author).

Article 30 then states that "Any person whose rights have been violated further to immatriculation may not assert a claim to property title. Solely in cases of fraud, a personal claim lies against the perpetrator of the fraud." Article 24 of the 2005 law establishes a special procedure which allows for transforming a *certificat foncier* into a *titre foncier*, which seems to imply that despite the land certification option, the title deed still offers more tenure security from a legal standpoint as the *titre foncier* contains non-rebuttable evidence of property rights, capable of defeating any competing claim based on a *certificat foncier* or customary use. However, one important exception exists which may yet play a crucial role in the determination of rights before the Malagasy courts. This is the tomb provision contained in Article 25:

> The land registry system applies to real property of any nature, whether including buildings or vacant. Tombs containing sepultures and their underlying land may be immatriculated, but they remain governed by the special property rules which govern them and which at all times, by their nature, can neither be sold, transferred, assigned and are exempt from execution or attachment.

Despite this provision, it is not impossible to imagine that competing claims to such ancestral lands (*tanindrazana*) can be asserted both by those who have established tombs when they first cleared the land and later migrants who took over the land and established their own tombs when the first comers moved away.

Chapter 5 of *the Loi N° 2006–031 du 24 novembre 2006 fixant le régime juridique de la propriété foncière non titrée* (the 2006 Act) governs the settlement of disputes. Article 24 states that any land disputes arising out of that law "fall under the exclusive jurisdiction of the Civil Courts". Article 25 however states that "The settlement of litigation or oppositions between individuals related to untitled real property must be first referred to the applicable conciliation and arbitration procedure in the relevant *collectivité*, prior to being referred to the relevant Court." In the areas where the land offices are operational such local grouping are referred to as local recognition commissions (see *infra*).

Clearly, there are questions to be settled even in the unlikely event that the PNF legislative scheme is effectively implemented in the short term. As stated, the argument for PNF is that it would bring customary landowners into the positive law system, but will Article 25 of the 2005 Act allow customary landowners to effectively resist challenges to their rights by using the tomb as their first line of defence? In short, PNF still seems to represent a perilous enterprise of trying to reconcile fluid customary arrangements to those of the requirements of positive law codifications.

### *How is Land Ownership Perceived at the Local Level?*

In order to be able to assess local customary arrangements of land access we will examine some local perceptions of ownership and the perceived legitimacy of land claims.

It is common practice in Madagascar that people who have cultivated a piece of land for generations still do not always get their rights to land recognized. As Kull states, "[l]egitimacy applies not only to institutions and leaders, but also to rules. In Madagascar, what is legal is often not legitimate; and what is legitimate is often not legal" (Kull 2002: 66, cf. Evers 2005).

Many local populations do not see the statutory property rights cited above concerning farmers who can lay claim to land on the basis of long-standing cultivation, but who have not themselves cleared the land, as legitimate (ibid.). As mentioned previously, in Madagascar, most people acquire land by way of inheritance. There exists a deep-rooted belief in the absolute land rights of the first cultivator of a land and his descendants. This perception is linked to the fact that Malagasy have built their tombs on the land of their ancestors (*tanindrazana*), a sign of family authenticity in the region.[10] Through tombs, Malagasy can have access to their ancestors, who play such an important role in their lives. One needs land in order to build a permanent tomb; the tomb is the portal to the hereafter and the entry point into the process of 'ancestralization,' a ritual sequence that facilitates the deceased return to the ancestors. Malagasy take great care in retaining positive relations with their ancestors; this translates into the daily practice of maintaining rich ritual and ceremonial lives. Losing one's land jeopardises these relations and can have serious repercussions on the living (cf. Evers 2002 and 2006). Many Malagasy even believe that, by losing land, people can become 'lost people' – *olona very*, a notion rooted in the country's history (Evers 2002, Graeber 2007).

Historically, dislocation from land has triggered violent conflicts between the state and local communities. During the pre-colonial Merina kingdom, when internal slavery was implemented on the island to meet

---

[10] It is important to note that land claims on the basis of tombs are not as straightforward as they might seem as it is not inconceivable that different waves of migrants who succeeded in gaining access to land and building their own tombs, might also lay claim to land ownership pointing to their family tombs.

specific political and economic objectives (Campbell 1991: 111), land was forcibly denied as a means of control, de-historicization and pacification, breaking links between Malagasy people and their ancestors and eventually making them *andevo* (slaves). *Andevo* were 'lost people,' lacking links to ancestral lands and permanent tombs and thus precluded from becoming ancestors themselves – the essence of the Malagasy identity (Bloch 1989; Evers 2002, 2006; Feeley-Harnik 1991; Graeber 2007). Colonial-era policies aimed at export production (see Sodikoff 2008) and forced labour regimes continued to break local links with the land, as many forests were appropriated for logging concessions (Jarosz 1993; Fremigacci 1978). Today, the term *andevo* still implies someone who lacks anchoring in land through tombs, land and history (Evers 2006: 417, Graeber 2007).

Relations with the ancestors are the essence of being Malagasy, entrenched in land practices and *fady* ('taboos') concerning inappropriate land management behaviour (including the sale of ancestral land to outsiders). On a daily basis, people work the land and renew their link with the ancestors, who in turn regenerate both people and land. This vital relation with the ancestors is reflected in the concept of *hasina* described by Delivré (1967: 167–84) as a form of energy innate to existence. References to *hasina* have been encountered by many scholars conducting research in Madagascar (cf. Bloch, 1989: 46–88, Delivré 1967, Dubois 1938, Edholm, 1971). Southall (1986: 414) translates *hasina* as "sacred ritual potency," and considers it to be a central concept for all groups in Madagascar: "Here is one of those pervasive themes which justify emphasis on the essential unity of all Malagasy culture, despite its apparent regional contrasts." Bloch (1989: 65) also writes that the notion of *hasina* is the "kernel of Malagasy thought." Fertility, successful harvests and good health are all ascribed to *hasina*. The belief that ancestors can activate destructive aspects of *hasina* causing, for example, infertility, illness and death, when people do not conform to envisaged ancestral expectations (see Cole 2001) is widespread in Madagascar.

In view of the above, in many settings access to land and lawful jurisdictions for the settlement of land disputes are often still determined within a context where the State and the court system exercise little influence. In such regions, the *tompon-tany* ("masters of the land")[11] hold decisional powers with respect to real property issues.

[11] The concept of *tompon-tany* exists throughout Madagascar. *Tompon-tany* is translated as 'master(s) of the land' and not 'land owner(s)' because even though most villagers

### *How People Evaluate Lex Fori Concepts*

It is precisely because Malagasy value their land in many more ways than just as a source of livelihood that implementation of PNF is complicated. Furthermore, as we over the years of research in various regions registered, Malagasy view the PNF as a *vazaha* import that tries to appropriate powers which the *tompon-tany* deem to be their sovereign prerogative. Local groups have trouble understanding the Western legal concepts being introduced wholesale into their lives. As the Land Observatory has noted: "What is more, the new terminologies introduced by modern law (tenure, right of way, contrary evidence, etc.) are as many concepts that the members of the local recognition commissions need to assimilate." (2009: 5). As will be explained later, *tompon-tany* usually play an important role in the commissions that advise the Local Land Office.

People often complain that the new land registration system is too complex to understand or is improperly implemented. That being said, outright vocal opposition has often been shied away from in Madagascar. Local groups and leaders are more likely to plead misunderstanding of the new system. (Van Dijk 2010: 54–55). But memories of the French colonial period and its appropriation of lands, forced labour and heavy taxation still seem to weigh on the collective mind. The fact that the PNF is inspired by the French Civil Code hardly favours universal acceptance in the short term. It is scarcely surprising that it has been labelled by some as a "colonial vestige" (cf. *Tany* 2011).

In addition, the implementation of PNF was made possible through financing by international donors using the technique of adopting local land offices. Whereas the law states that the LLO is established upon request of the mayor, NGOs frequently select areas of interest to them. The 2009 Malagasy Land Observatory report states: "While some donors complied with this process, others more heavily favored a project approach aiming to produce rapid quantitative results at the expense of the duration and quality of training, indispensable conditions for establishing true skills within the commune. In this case, the communes were selected without consulting the local authorities beforehand" (page 6). The question then is whose agendas are being realised with the implementation of PNF.

---

consider *tompon-tany* claims as legitimate, they rarely hold recognised land title deeds. They commonly claim to be the ones who first cleared the land for cultivation.

### *Whose Tenure Security?*

The 2009 report of the Malagasy Land Observatory discloses that: "In three years, the improved tenure security via the land certificate has accomplished the equivalent of 12% to 14% of what land registration achieved in 110 years" (ibid.: 4). But who is actually registering land? Fieldwork on the ground in the Manakara region in the context of the VU Amsterdam/University of Antananarivo collaboration and fieldwork by Evers in 2007 suggest that those deemed to hold customary ownership also were the most successful in obtaining their certificates with ease: *tompon-tany* and other local power holders. This is confirmed in a more recent study drawing on data of the Land Observatory, which neutralises the optimism in the 2009 report by adding that: "Certification rate is still rather limited (8.2% of the households and 5.6% of the plots) and certification is strongly determined by the households' wealth" (Burnod et al. 2012: 19).

We also noted that villagers who stood to derive the least direct benefit or with no relations to local power holders (both traditional and those in formal positions like the mayor) were also the most skeptical about the idea that the Land Registration Office would solidify their tenure status. Part of this suspicion may be historical, including more recent experiences with the workings of foreign implanted development projects and backroom deals that local power holders may have struck to facilitate such projects (cf. Kauw 2006, Van Dijk 2010). The Land Observatory recognizes the issue but believes that the answer lies in additional training of the people responsible for the land office, particularly the mayor, and the traditional power holders sitting on the local recognition commissions (2009: 5–7). This suggestion might be helpful but there is a fundamental conflict inherent in the training function. Training the mayor in project aims, concepts and terminologies might strengthen his role as a potential mediator between the project and the villagers but leaves him open to the charge that he speaks a "cultural idiom" that is foreign to the villagers.

Some Local Land Offices have been established without the prior consent of local power holders. Local groups find this an irritant, but this is even more problematic for the traditional local leaders like the *tompon-tany*. Trust and collaboration is a heady challenge in the face of such recent perceived lack of respect in their regard. On the other hand, in implementation processes where the traditional leaders were involved in the LLO from its inception, they might be tempted to see the potential advantages of the project. To stress the importance and influence of local leaders like the *tompon-tany*, they are usually invited by the mayor to sit on the Local Recognition Commission, whose nomination is confirmed by order of the *commune*. This committee advises the local land registration office on the local legitimacy of land registration requests (see also Malagasy Land Observatory November 2011 report on these procedures). In that role, the *tompon-tany* are likely to be more inclined to give credence to claims that are made by fellow *tompon-tany* rather than to renters of land who might have cultivated local plots of land for more than ten years or even generations. Indeed, the very notion of granting tenure to people who are *mpiavy* (migrants) could be considered anathema. Thus, *tompon-tany* land claims are easier to register as they only would have to find a modus operandi among themselves. Others in the village might also be reluctant to oppose such claims if they wish to keep on living in the village.

### *The Status of Women as Landholders*

Gender relations are an important aspect of village social hierarchies, which necessarily impact on land certification, practices. The PNF legislation is touted as being gender neutral. The Land Observatory mentions that 14.403 women (Malagasy Land Observatory website 13 December 2012) have obtained a land certificate. Most NGO's financing the Land Registration offices clearly target women as beneficiaries of the projects. There is a logic to this in the fight against poverty, as women who head households are deemed to be the most vulnerable for living in poverty as the study of Stifela et al. demonstrates: "we found that female headed households suffered 6–14% lower consumption." (2010: 594). The question arises as to which women would benefit from land registration.

In many Malagasy rural settings, land ownership of women is something that is somewhat problematic (Burnod et al. 2012: 16), particularly in the Manakara region where we followed land registration projects over a period of five years. Patrilineal heritage systems made men the heirs and cases of women actually inheriting land were rare (Kauw 2006: 53). They were usually widowed women who theoretically were under the legal duty to return the land back to their family but were on occasion offered the option to continue cultivating the land and passing it onwards to their sons. Actually registering such a plot of land is a potentially dangerous enterprise. During 2007 fieldwork in the Manakara region, two cases where women obtained a certificate were brought to our attention. One of these women was said to be poisoned by her in-laws and the other was socially ostracised and expelled from the village. She left the countryside altogether and survived with her children begging in the streets of Manakara.

In 2010, Aliene van Dijk (page 49) noted that by March that year, 57 women in the Manakara region had obtained a land certificate. They appeared to be members of well-to-do families, women with better education and presumably with more knowledge of PNF. Also, it is not to be excluded that social links with the mayor and other people of authority might have facilitated granting of the certificate. Possession of the certificate was no guarantee that they were free to determine the fate of their land in terms of management or selling it, as they remained subject to social constraints imposed by the local cultural system of authority of (elderly) males. The above listed predicaments of women that we noted in our research are confirmed in a study of the Land Observatory (June 2011), which states that while according to *lex fori* legislation women should have equal access and rights to land, in reality *lex loci* customary arrangements prevent them from claiming such rights (page 39).

### *Conflict Resolution*

In the first instance, the local *tompon-tany* might find it in their interest to collaborate with the land registration project. As some *tompon-tany* during our studies pointed out: "Yes, the state is behind us and confirms that we are the rulers and owners of the land in this region" (data fieldwork Evers 2007). However, by the mere act of registering, *tompon-tany* may have acceded, or in legal parlance, attorned to the jurisdiction of the courts. Before land registration, local leaders settled land conflict at the local level. In many regions authenticity demonstrated through the family tomb of a particular family was an important factor in assessing conflicts (the domain of *lex loci*). With the new system *tompon-tany* theoretically can lose their land, for example by way of mortgage foreclosure, or through the large-scale land acquisitions increasingly common in Madagascar and elsewhere in Africa (see Seagle and Vinciguerra in this volume). And under Articles 29 and 30 of the 2005 Act, even where a complainant proves to the satisfaction of the courts that such land was obtained through fraud, the plaintiff would receive a money award, but not necessarily return of his land.

So in the short term, the PNF allows a say in securing substantive property rights because the *tompon-tany* themselves often make up the review body (local recognition commissions) which assesses the validity of claims concerning land certification. However, once confirmed, the courts will become the exclusive arbiter of dispute resolution in the subject matter if it cannot be locally resolved. This change of forum is capital. The transfer of jurisdiction risks undermining the supremacy of the *tompon-tany* and a shift towards a positive law system and everything which that implies – taxes, conveyance of land, granting of mortgages, liens and similar security interests, notwithstanding the consolidation of power of the *tompon-tany* in the short term. As the study of Stifela et al. on the evolution of groupwise poverty in Madagascar between 1999–2005 demonstrates, the land rich flourish in remote areas, and benefit more from land investments and economic opportunities (2010: 595). Thus, while potentially bringing tenure security for some, PNF might contribute towards elite appropriation of the system and trigger deeper divisions within communities.

Customary arrangements of land tenure undoubtedly give rise to disputes and intensification of local hierarchies. PNF, however, has it own challenges and might actually increase local land conflicts and social cleavages. The Land Observatory also has observed that PNF has failed to provide a proper forum for such conflicts. In theory, people with land certificates engaged in a land conflict can seek a solution in the Malagasy courts, but villagers often view the courts as corrupt, procedures as cumbersome and complex and costs as prohibitive. However the Land Observatory report (2009) quotes a deputy mayor in the *commune* Amparafaravola district (Alaotra – Mangoro region), stating that the number of land conflicts for which arbitration was sought at the *commune* level had almost tripled since creation of the Land Office in October 2007: "The opening of the Local Land Office (LLO) pushed inhabitants to bring forth latent conflicts that were real but that they had attempted until then to resolve more or less clumsily within the family or neighborhood" (Land Observatory 2009: 5). However, the report then mentioned that "most LLO prefer not to process contentious requests and do not wish to take on the role of proximity arbitration body" (ibid. 6). Thus PNF does not offer efficient and effective conflict resolution mechanisms concerning land matters.

### *Development Means Taxation*

The causal link that Hernando de Soto (2000) predicted between land registration and agricultural investment (via bank loans using the land as collateral) has yet to materialize. As Jacoby and Minten (2005: 15) point out "agricultural lending as well as the establishment of a legal framework for mortgaging land, [are] conditions which are not generally present in Sub-Saharan Africa." Also farmers are often not deemed credit-worthy by banks (Otto 2009: 188). The increase in investment and production are not realistic results of land registration in the African context, as Jacoby and Minten (ibid.: 1) substantiate in their Malagasy study, they found "no significant effect of having a [land] title on plot-specific investments and corresponding small impacts on land productivity and land values". Lack of money for investment is not the only factor crucial in understanding Malagasy expenditure patterns concerning land; for them land actually has not just an economic value. Other non-economic factors, as argued above, are at the heart of farmers' decision-making patterns concerning their: 1) ability to register land; 2) willingness to register land; 3) allocation of their budget to land. For the same reason, even the hope of the Malagasy government and its international donors that registration would pave the way for taxation of the land turns out to be a hazardous route, precisely due to the fact that Malagasy conceptualize and value their land very differently from the international stakeholders involved in the land registration projects.

PNF documentation clearly links the programme to the 'development' of local communities through increased productivity and investment by using the land as collateral. The taxation system in theory should bring 'development' to the villages. Even in the literature of the land observatory 'development' is presented as the result of taxation as the benefits of the new taxation system would be used for local communities: "Among other things, the Land Policy's ultimate goal is land tenure management conducive to the development of decentralized territorial governments via the provision of territorial management and taxation tools." (2009: 3). How such money precisely should be allocated and what 'development' would entail is not detailed.

With regard to decentralisation of taxes, the Land Observatory report does disclose a jurisdictional tangle, which if left unresolved is sure to give rise to disputes. While land registration of certificates is decentralised under the PNF, the national land title system is still in service as well and there appears to be a dearth of information flow between the two systems. In the words of the Land Observatory: "Yet, maintaining centralized land management is a complete mismatch with a decentralized land taxation system. Unlike land certificates, the information on land titles does not go through the communes; the communes do not know the status, occupation, and size of titled lands, or even the holders of taxable rights."

(ibid.: 8) In a November 2011 assessment study of the PNF the lack of a proper taxation system was again deplored by the Land Observatory (page 39). Even presuming a functional taxation system, the willingness of farmers to pay outstanding taxes is far from assured, as they might buttress their claims to land with reference to tombs and ancestors. As a *tompon-tany* landowner in the Ambalavao region explained during 2007 fieldwork:

> You think we are going to pay taxes on our own land to the State? This is the land of our ancestors. We will never pay money for anything that already belongs to us. This is surely a *vazaha* idea. Just like before, they are coming to steal our money and our land. I warned the other *tompon-tany* to take measures so *vazaha* do not come to steal our land. After that they will want us to work as *andevo* for them. That never!

The Land Observatory has contemplated a series of measures, which they refer to as user incentives, designed to encourage the payment of property taxes: "temporary tax exemptions for certified lands, massive reduction of the cost of certificates during a set period of time, etc. On the contrary, unregistered land could be subject to higher taxation." (2009: 15) Based upon our research and the studies of the Land Observatory (see November 2011 report), we suspect that the *tompon-tany* quoted above is not alone in his views on the reformed taxation system. His view that *vazaha*, foreigners, might come to steal their land is not uncommon in Madagascar, a belief anchored in both the land confiscation practices of the colonial past and more recent phenomena.

### *When the Foreigners Come*

Despite the division between State land and 'untitled private property' (derived from 2005 legislation), PNF potentially can give rise to a phenomenon that could make the worst fears of Malagasy come true: the prospect of losing one's land to third party interests, whether from inside Madagascar or from foreign interests, thereby severing the link with the ancestors. The state might indeed label land that is not registered as open for foreign investment and lease it to foreign companies, who commonly enter the market through Malagasy subsidiaries (Evers et al. 2011).

Ravalomanana managed to increase national growth figures while leasing land to foreign companies for mining and agricultural purposes (Ploch 2011: 11). Simultaneously, conservation NGO's continued to co-manage Malagasy rainforests (including delegated management structures like the Wildlife Conservation Society in Masoala). Even going so far as insisting on legislative changes to *tavy* practices, while stimulating land registration and levying penalties against trespassing in 'national' parks (Kull & Laris 2009). Mining companies have been granted access to local grounds with the discursive legitimization that these lands are 'unused' or 'unproductive' while local farmers are portrayed as ignorant and desperately poor. Such land deals are often framed as a 'development opportunity' for the state and local communities, with promises of education, 'economic growth' (and thus improved livelihoods), technology, infrastructure, education and healthcare, and alternate income-generating activities (e.g. employment on concessions, ecotourism) increasingly seen as 'trade-offs' (or compensation) for the lease or sale of large swaths of arable land (Smaller, Carin and Mann, Howard 2009; IIED and FAO 2008, Seagle in this volume).

Ravalomanana's economic agenda came to a grinding halt when news became public of the South Korean company Daewoo's proposed 99-year lease of 1,3 million hectares of land meant for agricultural production. It sparked massive protests and political dissent within Madagascar, providing a platform for the protests led by Andry Rajoelina in 2009, which culminated in a *coup d'état* (see also Vinciguerra in this volume). A reason why Rajoelina had been so successful in mobilising Malagasy in support originates in his rhetoric on foreign investors like Daewoo's attempts to destroy local land based livelihoods by the claiming of ancestral lands, thus harnessing what is probably one of the greatest fear of many Malagasy.

The protest against the Daewoo case eloquently illustrates the extent to which land matters in Madagascar are anything but a trivial matter. The obvious example is the tomb. Which indeed can be used as a pervasive rhetorical strategy to assert claims in land competition. From this standpoint, the tomb is then presented as the quintessence of interconnectedness, a medium intimately related to kinship networks and heritage. Its very existence is dependent upon possession of land, and the tomb as symbol and icon is emblematic of existence itself – both socially and ontologically, so powerful as medium, that it is able to sweep aside *lex fori* notions of ownership. The reference to tombs in the legislation does little to safeguard villagers from losing their ancestral land when the state claims the land for purposes other than subsistence farming. The prospect of fighting this in court is scarcely an option for most farmers.

The political crisis had enormous consequences for Madagascar as international aid programmes of NGO's, the United States and the European community were initially put on hold and spending strategies reconsidered. As the Word Bank Madagascar country brief highlights: "Large fraction of official aid – which represents 40 percent of the budget and 75 percent of the public investment program – remains on hold. That has led to significant cuts in investments, a decline in the delivery of social services and further deterioration of growth." (World Bank 2012). Madagascar remains highly depended on outside donors. The Land Observatory also observes this dependence and already in a 2009 urged donors to continue their institutional, technical and financial support for the land titling programmes. Currently, the political crisis is ongoing with no solutions on the horizon. In the meantime, Malagasy poverty continues to be endemic, and optimism for a better future is fading. Madagascar's poverty ratings register increased poverty levels by above nine percentage points between 2005 and 2010, reaching 77 percent of households, the highest rate in Africa (World Development Indicators, 2011). Many Malagasy are driven back into farming (Stifela et al. 2010: 31), renewing the pressure on land and potentially exacerbating conflict at the local level.

## *Conclusion*

The central aim of the PNF holds out the promise of uniformity, predictability and that it ultimately will reduce tensions and disputes in the field of land access and ownership. The PNF, however, at this point suffers from three major structural impediments. First of all, the system is widely misunderstood. Secondly, the system is only partially implemented and it remains to be seen whether it is being implemented in the same manner from one locality to the next. Thirdly, the PNF is not enforced at this point.

Added to these three structural difficulties is the multitude of actors and stakeholders to land at the national, local and international levels, ranging from village elders to conservation groups to mining companies to members of the Malagasy diaspora. The recent introduction of mega-land transactions into the landscape creates another variable impacting on power relations within the country.

Finally, and crucially, PNF is far from universally accepted. Although presented as the comprehensive, watertight system that will finally bring certainty to land relations, in all likelihood the contrary result can be expected. In fact, if one were to ask the question: what do Malagasy agree upon generally, the answer most likely would not be the PNF, but quite something else. For there are two things that are still used as powerful discourses by Malagasy laying claim to land, elements that they deem to be universal Malagasy tenets when it comes to land: 1) tombs and the underlying land are sacred. 2) land comes from the ancestors. Any violation of these fundamental beliefs will bring ill fortune to those who are responsible for the alienation of land from the ancestors.

## *Bibliography*

Bloch, M. 1989. *The Disconnection between Power and Rank as a Process: an Outline of the Development of Kingdoms in Central Madagascar.* In: Bloch, M. (ed.), Ritual, History and Power, Selected Papers in Anthropology, pp. 46–88. London: The Athlone Press.

Burnod, P., N. Andrianirina, C. Boué, F. Gubert, N. Rakoto-Tiana, J. Vaillant, R. Rabeantoandro, R. Ratovoarinony. 2012. Land reform and certification in Madagascar: Does perception of tenure security matter? Word Bank conference (23–26 April 2012) http://www.landandpoverty.com/agenda/pdfs/paper/burnod_full_paper.pdf (accessed 13 December 2012).

Campbell, G. 1991. The State and Pre-colonial Demographic History: The Case of Nineteenth-century Madagascar. *The Journal of African History* 32(3): 415–445.

Cole, J. 2001. *Forget Colonialism? Sacrifice and the Art of Memory in Madagascar.* Berkeley and Los Angeles: University of California Press.

Delivré, A. 1967. *Interprétation d'une tradition orale. L'Histoire des Rois d'Imerina.* Unpublished thesis, Paris. Published in 1974 under the same title. Paris: Klincksieck.

Dijk, A. van. 2010. Staking Out Your Turf: An investigation into a Land Registration Project in South East Madagascar. Unpublished master's thesis VU University Amsterdam.

Dubois, H.M. 1938. *Monographie des Betsileo.* Paris: Institut d'Ethnologie.

ECR. 2006. *Les données de référence sur les coûts et temps moyens requis pour l'exécution des procédures foncières.* Antananarivo: Millenium Challenge Account.

ECR. 2008. *Les données de référence sur les coûts et temps moyens requis pour l'exécution des procédures foncière.* Antananarivo: Millenium Challenge Account.

Edholm, F. 1971. *Royal Funerary Rituals among the Betsileo of Madagascar.* Ph.D. diss, London School of Economics.

Evers, S.J.T.M., P. Burnod, R. Andrianirina Ratsialonana, A. Teyssier. 2011. Foreign land acquisitions in Madagascar: Competing jurisdictions of access claims. In T. Dietz, K. Havnevik, M. Kaag, T. Oestigaard (Eds.), *African engagements: Africa negotiating an emerging multipolar world* (Africa-Europe Group for Interdisciplinary Studies Series 7) (pp. 110–132). Leiden & Boston: Brill.

Evers, S.J.T.M. 2006. Expropriated From the Hereafter: The Fate of the Landless in the Southern Highlands of Madagascar. *Journal of Peasant Studies* 33, 3: 413–444.

Evers, S.J.T.M. 2005. Trumping the Ancestors: The Challenges of Implementing a Land Registration System in Madagascar, *Competing Jurisdictions: Settling Land Claims in Africa,* 2005, Sandra J.T.M. Evers, Marja J. Spierenburg, Harry Wels (eds.), pp. 223–242, Leiden: Brill Academic Publishers.

Evers, S.J.T.M. 2002. *Constructing History, Culture and Inequality. The Betsileo in the Extreme Southern Highlands of Madagascar.* Leiden: Brill Academic Publishers.

Evers, S.J.T.M. and Seagle, C. 2012. Stealing the Sacred: Why "Global Heritage" Discourse is Perceived as a Frontal Attack on Local Heritage-making in Madagascar. *Madagascar Conservation & Development* (S2): 97–107.

Fremigacci, J. 1978. *L'administration coloniale: les aspects oppressifs. Omaly sy Anio* (Hier et Aujourd'hui) Revue d'Etudes Historiques Antananarivo 7–8: 209–237.
Graeber, D. 2007. *Lost People. Magic and the Legacy of Slavery in Madagascar*. Bloomington: Indiana University Press.
IIED and FAO. 2008. *Fuelling Exclusion? The Biofuels Boom and Poor People's Access to Land.*
Jacoby, H. and Bart M. 2005. Is Land Titling in Sub-Saharan Africa Cost-Effective? Evidence from Madagascar. Paper part of a larger study entitled "Land Titles, Investment, and Agricultural Productivity in Madagascar". (retrieved on 18 August 2011: http://siteresources.worldbank.org/DEC/Resources/LandTitles_Mad.pdf).
Jarosz, L. 1993. Defining and Explaining Tropical Deforestation: Shifting Cultivation and Population Growth in Colonial Madagascar (1896–1940). *Economic Geography* 69(4): 366–379.
Kasanga, K. and Woodman, G.R. 2004. Ghana: Local Law Making and Land Conversion in Kumasi, Asante, in G.R. Woodman, U. Wanitzek, H. Sippel (eds.), *Local Land Law and Globalization,* Münster: LIT Verlag, pp. 153–216.
Kauw, K. 2006. *Decentralization and Land Registration in a Nascent Democracy – The Case of a Rural Community in South East Madagascar*. Unpublished Master's thesis VU Amsterdam.
Kull, C. 2004. *Isle of Fire: The Political Ecology of Landscape Burning in Madagascar*. Chicago: Chicago University Press.
Kull, C.A. and Laris, P. 2009. Fire ecology and fire politics in Mali and Madagascar, pp. 71–226 in Mark A. Cochrane, ed. *Tropical Fire Ecology: Climate Change, Land Use and Ecosystem Dynamics*. Springer-Praxis, Heidelberg, Germany.
Malagasy Land Observatory (*Observatoire du Foncier*). 2009. The Local Land Offices: Commune Service or Implants Set by Development Projects? *Land Scope* (*Notes du l'Observatoire du Foncier.* Nr. 3, November 2009.
Malagasy Land Observatory (*Observatoire du Foncier*). 2011 (June). Propositions d'amélioration de la politique et des stratégies dans la réforme foncière. Enquêtes menées dans quatre Régions de Madagascar.
Malagasy Land Observatory (*Observatoire du Foncier*). 2011 (November). Evaluation de la réforme foncière à Madagascar: Rapport final synthétique.
Malagasy Land Observatory (*Observatoire du Foncier*). 13 December 2012 http://www.observatoire-foncier.mg/.
Merlet, P. and Bastiaensen. J. 2012. Struggles over property rights in the context of large-scale transnational land acquisitions. Using legal pluralism to re-politicize the debate.
Illustrated with case studies from Madagascar and Ghana. Discussion Paper May 2012 (ISSN 2033–7329). University of Antwerp.
Otto, J.M. 2009. Rule of Law Promotion, Land Tenure and Poverty Alleviation: Questioning the Assumptions of Hernando de Soto, *Hague Journal on the Rule of Law*, vol. 1, nr. 1, pp. 173–195.
Ploch, L. 2011. Madagascar's Political Crisis. Washington: Congressional Research Service.
Pronk, C. and S.J.T.M. Evers. 2007. The complexities of land access in Southeast Madagascar. *Taloha, revue scientifique internationale des civilisations* 43(18).
Razafiarijaona, J. 2009. Bilan diagnostic de sécurisation foncière et perspective pour Madagascar. *Revue Juridique et Fiscale de MCI*, n. 45.
Smaller, C and Mann, H. 2009. A Thirst for Distant Lands: Foreign Investment in Agricultural Land and Water, International Institute for Sustainable Development (IISD) (May 2009).
Sodikoff, G. 2007. An Exceptional Strike: A Micro-history of 'People versus Park' in Madagascar. *Journal of Political Ecology* 14: 10–33.
Soto, de, H. 2000. *The Mystery of Capital: Why Capitalism Triumphs in the West and Fails Everywhere Else*. New York: Basic Books.
Stifela, D., Forster, F. and Barrett, C.B. 2010. The Evolution of Groupwise Poverty in Madagascar, 1999–2005. *Journal of African Economies*, Vol. 19, number 4, pp. 559–604.

*Tany*, Le Collectif pour la Defense des Terres Malgaches. 2011 (retrieved 18 August 2011 http://terresmalgaches.info/spip.php?article40).
Teyssier, André. 2010. Decentralizing Land Management. The Experience of Madagascar. *Perspective*, 4, Juin 2010.
Teyssier, A. 2004. *Programme National Foncier, contribution à l'élaboration d'une politique publique de sécurisation des droits sur le sol,* Direction des Domaines et des Services Fonciers, Propos d'étape nr, 1, période mars/juin 2004, Antananarivo.
Teyssier, A. Ramarojohn, L., Andrianirina Ratsialonana, R. (2010) Des terres pour l'agro-industrie internationale ? Un dilemme pour la politique foncière malgache *Echogéo,* 11, décembre 2009/ février 2010. http://echogeo.revues.org/index11649.html.
Teyssier A., Andrianiriana Ratsialonana R., Razafindralambo R., Razafindrakoto Y. 2009. Décentralisation de la gestion des terres à Madagascar : processus, enjeux et perspectives d'une nouvelle politique foncière. In Colin J.-P. et al. *Les politiques de reconnaissance des droits fonciers. Du cadre légal aux pratiques locales*. IRD Karthala : 273–297.
World Bank, Data and Statistics. 2012 (retrieved 13 December 2012): http://databank.worldbank.org/ddp/home.do?Step=3&id=4.
World Bank, Madagascar Country Brief. 2012 (retrieved 15 December 2012): http://web.worldbank.org/WBSITE/EXTERNAL/COUNTRIES/AFRICAEXT/MADAGASCAREXTN/0,,menuPK:356362~pagePK:141132~piPK:141107~theSitePK:356352,00.html.
World Development indicators. 2011. (retrieved 15 December 2012): http://data.worldbank.org/country/madagascar.

# LAND RIGHTS AND ALIEN PLANTS IN DRYLAND MADAGASCAR

Karen Middleton[1]

Over the past twenty years the Government of Madagascar has signed up to multiple initiatives in the fields of development and environment, including the 1992 UNEP Convention on Biological Diversity, the Millennium Development Goals (2000), and the tripling of land under protection for biodiversity (Durban Presidential Declaration 2003). Land titling has been consistently presented as a key adjunct of these programmes. Advocated as the solution to the island's perceived problems more or less since the first days of the former French colonial administration, arguments for formalizing land rights enjoyed a particular resurgence in the 1990s when Madagascar's increasing poverty, together with its emergence as a "mega[bio]diversity country" (Mittermeier 1988; Myers 1986), made the integration of development and environmental concerns a stated priority for key international funders, including the World Bank and USAID. It was argued that land titling would regularize the boundaries of protected areas and gazetted forests while providing peasant farmers with greater incentives to invest in their land. Investment in tree crops, particularly in fruit trees, was among the on-farm improvements that better security of tenure was expected to encourage Malagasy peasantry to undertake (World Bank 1990: 5).

In fact, the presumed correlation between formal title and improved resource management that underlies arguments for a land titling programmes is far from proven. The global literature has failed to demonstrate either that the extension of private land ownership is necessarily followed by agricultural intensification or that the adoption of soil conservation measures requires formal registration of land. Thus, one recent

[1] Fieldwork in southern Madagascar was supported by the Economic and Social Research Council (UK), the Leverhulme Trust, and the Nuffield Foundation. In Madagascar I should like to record my gratitude to the *Ministère de la recherche scientifique et technologique pour le développement* and to staff of Cap Sainte Marie Réserve Spéciale. Thanks are also due to Sandra Evers, Sam Middleton, Philip Thomas, and two anonymous referees for valuable comments on earlier drafts; all errors and shortcomings remain my own. As ever, my greatest debt is to the peoples of Karembola for their generosity and patience.

study of an intensively titled rice-growing area in the Madagascar highlands found that having a title had no significant effect on plot-specific investment and correspondingly little effect on productivity (Jacoby and Minten 2007). Conversely, formal registration of land proved to be unnecessary in the now classic case of Akamba agro-pastoralists in Machakos District, Kenya who over a period of sixty years reversed a trend of land degradation, conserved and enhanced tree stocks, and improved overall productivity. Here peasant confidence to invest in their farms was secured by "strong [individualised] ownership rights" under customary law (Mortimore 1998).

Contributors on both sides of the debate around land titling in Madagascar often frame their discussion in terms of simple, dualistic contrasts between a "modern" system of formal land registration introduced into the island by the former French colonial administration and "customary" or "traditional" tenure practices (Keck et al. 1994; Schoonmaker-Freudenberger 1995; Pronk and Evers 2007; Evers chapter 6 this volume; cf. McConnell 2002). Such an approach is unhelpful since state law in Madagascar very early recognised customary rights inherent in settled cultivation while the practice of so-called "traditional" or "customary" land tenure in turn has been strongly modified by historical contingencies, including state legislation and state decisions that have persistently tied land rights to cleared and cultivated land.

The tendency to cast the problem in terms of binary oppositions *between* two systems of land ownership – one based on "custom" (*lex loci*) and the other on "positive land law" (*lex fori*) – has two further consequences. Firstly, it is associated with a proclivity to discuss customary land tenure in terms of a putative generic "Malagasy", whose perceptions of ownership and of the legitimacy of land claims are informed by a relatively homogenous corpus of beliefs about the ancestors, the elders, and spiritual power. Largely extrapolated from Merina, Betsileo and Betsimisaraka ethnography, such an approach not only glosses the actual diversity of beliefs and tenurial practices to be found in the island but also risks obscuring contestation within local communities. Secondly, an overly dichotomous approach means that we generally lack detail on how customary proprietary rights have been recognised and reshaped by agents of the state. Thus, one overview of land rights issues in Madagascar highlights a conflict between whether a piece of land is legally held according to written law or legitimately occupied according to verbal accounts (Healy and Ratsimbarison 1998: 296–7). Such simple contrasts between writing and orality, and between European law and Malagasy tradition, are unhelpful because even in Europe, held to be the *fons et origo* of written law in Europe's former colonies, alternative proofs are used to determine land entitlement when written documentation is deficient. For instance, in British courts reference may be made to trees or hedging in the judgement of boundary disputes. One might suppose that trees and hedging would figure even more prominently in tenurial arrangements in Madagascar, where so few rural citizens have paper title, and where the state has regularly recognised planted tree stock as evidence of occupation ("ownership") of land, yet very little has been written on tree tenure in Madagascar compared to the depth of ethnographic study elsewhere in the developing world (Fortmann and Bruce 1988; Raintree 1987).[2]

This chapter reflects on land tenure practice and land tenure issues in two farming communities in the Karembola region of dryland southern Madagascar. Both communities consisted of subsistence farmers who were directly dependent for their livelihoods on the land. Most land was untitled and technically belonged to the state. Yet people in both communities spoke frequently of land they gardened or otherwise exploited using concepts such as "owners" (*tompo*) and "inheritance" (*lovañe*). Exploring the tensions *and* the convergences between state interventions and local constructs of custom that inform smallholder decision-making, the chapter suggests that, rather than phrase the problem of land tenure in Madagascar in terms of competing jurisdictions, it might be more helpful to examine "traditional" or "customary" law and "modern" state law as co-evolutionary over time.

The broader objective is to highlight the complexity of factors to be considered when modelling correlations between agricultural intensification, biodiversity conservation, and land titling. Neither case study is "typical" of land tenure issues in Karembola, let alone in the "spiny dry forest" region of southern Madagascar more generally. On the contrary, the aim of the comparative ethnography is partly to query the appropriateness of "one size fits all" solutions by highlighting significant local diversity. To provide sharper definition to the argument about locality, the analysis

2 Philip Thomas explores how the adoption of colonial perennials (coffee, eucalyptus, cloves) by farmers in Manambondro (Southeast Madagascar) extended 'private' ownership by allowing new forms of boundary making (Thomas 1996, 27–29) but provides less detail on how this development interacted with state tenurial law. For the Highlands Alain Bertrand describes a pioneer front involving eucalyptus associated with collective land title (Bertrand 2004) but see Teyssier 1998 for another interpretation.

focuses on the changing role of an introduced plant species in validating proprietary claims. This focus serves both to highlight conflicts of interpretation and practice within the communities studied and to underscore the importance of looking at mutually shaping interactions between so-called "custom" and the "state".

## *Prickly Pears and Land Title*

Prickly pear is the common English name for *Opuntia* (Cactaceae), a Native American family containing scores of varieties. Prickly pears (or *raketa* as they are known in Malagasy vernacular) have been grown in dryland Madagascar for over two centuries. Well adapted to the low and irregular rainfall, the fruits and leaf pads have been harvested for human consumption, the leaf pads have been exploited to feed and water livestock, and the spinier varieties have provided stout hedging for garden plots and cattle pens (Defoort 1913: 153–4, 180, 183–5, 207, 233, 235; Decary 1930: 125–8, 144, 157, 181–2, 205–7). Prickly pears have also been planted to counter wind and soil erosion. As a highly productive perennial offering poorly capitalized farmers a cheap and easy form of agricultural intensification and soil conservation (cf. Barbera et al. 1995; Le Houérou 1996), Opuntia are from many angles just the kind of tree crop that international donors might hope land titling would encourage Madagascar's peasantry to plant.

On the other hand, some varieties of prickly pear have become invasive in introduced habitats. In the twentieth century the perceived costs to agriculture and stock-farming led to chemical and biological control of certain species in Madagascar, Australia, India, South Africa and elsewhere. Not that it is always a simple matter of distinguishing "good", "productive" varieties from "bad", "invasive" ones. In the 1920s French colonisers in Madagascar were divided over plans to eradicate the most common variety in the southern parts of the island, some viewing it as an economic plant and others as a weed (Middleton 1999; cf. Beinart and Middleton 2004).

When in the 1920s this particular prickly pear was decimated by an introduced cochineal insect, the colonial administration concluded that the outcomes had not been altogether beneficial for Madagascar's Deep South and accordingly initiated various Opuntia planting schemes in the region for fodder production, soil conservation, and as a human food. The most ambitious of these initiatives was *Operation raquettes*, launched in the late 1940s with post-war FIDES funding. But the most significant from the perspective of this chapter was the decision taken in the early years of the First Republic (1960–1972) to recognize cactus-planting as a form of reforestation and a specified way for men in the Madagascar drylands to discharge their statutory obligation to plant trees. Under Ordonnance 62–093 passed in October 1962 and as amended in December 1963 by Law 63–032, all able-bodied males in Madagascar aged between twenty-one and fifty-five years were required to participate in the work of national reforestation (*l'oeuvre nationale de reboisement*) by planting at least fifty forest or fruit trees annually or by contributing to forest protection or soil conservation work. Any man unable to discharge this national duty in person might get it executed by a third party or pay 100 francs in compensation each year.[3] Following a petition submitted to the National Assembly by Deputies representing constituencies in the region, the rules governing the application of this law were subsequently amended to allow men in Karembola and other parts of the Malagasy Deep South to discharge their civic duty by planting prickly pears.

The important issue of how the tree-planting legislation of the 1960s interacted with the broader corpus of land use and land tenure law that was passed under President Tsiranana has not been addressed by extant scholarship (Phillips 1986; Randrianarijaona 1983; Gade & Perkins-Belgram 1986; Kull 1996). But I would argue that one consequence of their interaction was that prickly pear plantings became the recognised way for Karembola farmers who lacked paper evidence of title to "document" entitlement in plots of land when land disputes were referred for resolution to agents of the state (cf. Middleton 2013). At the same time, it is important to note that multiple references in colonial literature to prickly pear hedging indicate that these plants had long been used to "parcel up" land in Karembola, forming an intrinsic component of "customary land tenure law". Thus, the decision of Madagascar's first post-independence government to recognize prickly pear as "fruit trees" in the context of the Malagasy Deep South illustrates the point that state legislation, far from implementing singular laws ("positive land law" [*lex fori*]) all over the whole island, has been regularly compromised by regional histories and ideas derived from region-specific "customary practice" (*lex loci*).

[3] *Ordonnance* 62–093 of 1 October 1962 and its *Décret d'application* as amended by *Loi* 63–032 *sur l'obligation au reboisement, devoir national*, 24 December 1963 (*Journal Officiel de la République Malgache* 79(329) 38 décembre 1963 p. 2808).

Advocacy of *Opuntia* as a "development" option for the Karembola region continued in the early years of Malagasy socialism, though little state-sponsored planting actually took place. Since the early 1990s, however, the public argument has been increasingly made by conservationists that introduced Opuntia are invading the indigenous spiny dry forest, threatening unique biodiversity of global value (Rauh 1995: 55, 68; 1998, xi, 66; Soarimalala and Raherilalao 2008: 159). The last twenty years or so has in effect seen a dramatic revision of conservation priorities in the island. International opinion which formerly focused on threats to Madagascar's rainforests began to argue that the native biodiversity of the southern drylands was seriously underrepresented in national networks of protected areas (DuPuy and Moat 1996; Olson and Dinerstein 2002: 203). Concurrently, this region, where deforestation was once reputed by certain commentators to be the slowest in Madagascar, is now alleged to be suffering some of the island's fastest rates (Sussman et al. 1994; Sussman et al. 2003; World Bank 2003; MEFT et al. 2009; but see Clark et al. 1998; Bodin et al. 2006; Elmqvist et al. 2007; Tengö et al. 2007).

### *"Show the* Vazaha *your Prickly Pears!"*

The first set of data to be discussed in this chapter was collected in Anjedava village, in Firaisana Marolinta, in 1991 (see Figure 1). Anjedava is one in a chain of Hazohandatse settlements strung out along the escarpment to the east of the Menarandra River where they straddle the "red soils" of the plateau and the sands to the south. It would be inaccurate to describe villagers here (or indeed at Cap Sainte Marie [Tanjona Vohimena], the second case study) as pastoralists because all were subsistence farmers growing a range of field crops (manioc, maize, sweet potatoes, millet, pumpkins, melons and various legumes). A few devoted some land to cash crops notably groundnut, and many more sold minor food surpluses in good years; but productivity was limited by hand hoe technology, and household production slipped easily into food deficits in bad years. It would be equally inaccurate to gloss villagers in either community as agro-pastoralists. To be sure, cattle were the quintessential sacrificial beast, the stipulated prestation in ceremonial exchange, the indicator of status, and the store of wealth, particularly against "bad years" when they were sold in a falling market to raise income to buy in food. However, livestock holdings varied immensely between households in both communities, with a few households owning in excess of 100 cattle and "innumerable" goats while others possessed none at all.

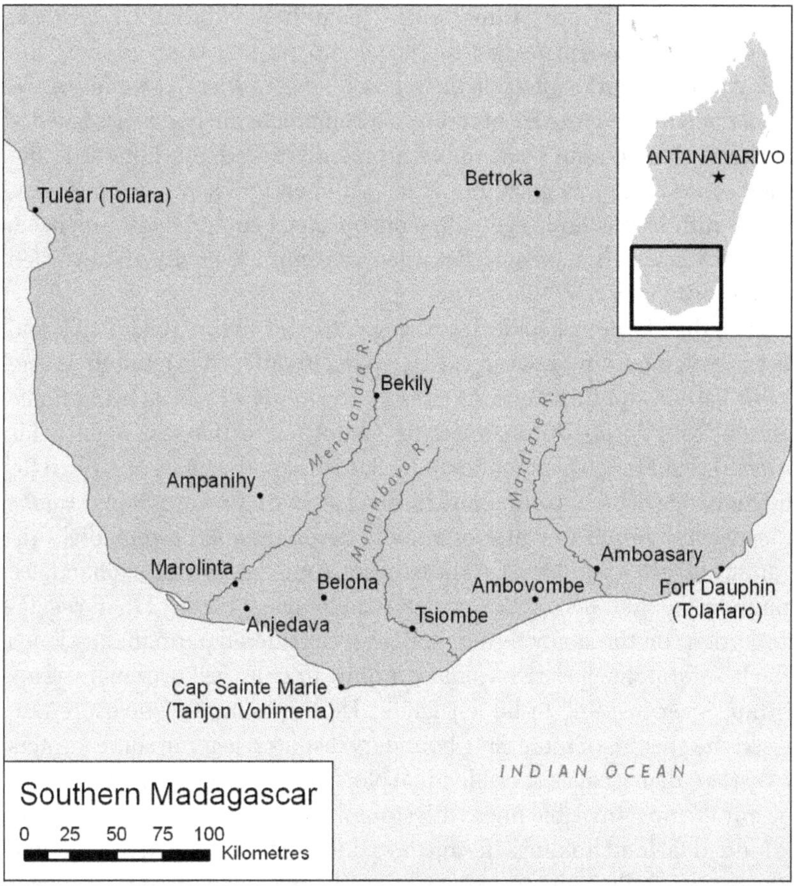


Figure 1. Southern Madagascar, showing the locations of the two field study sites discussed in this essay, Anjedava and Cap Sainte Marie (Tanjon Vohimena). Drawn by Michael Athanson, Map Librarian, Bodleian Library, University of Oxford.

In 1991 most inhabitants of Anjedava reported that droughts were becoming more frequent, escalating incidents of food insecurity, prompting important changes in land use, and stimulating conflict over land. The first and most significant of the reported changes in land use was that farmers had been relocating their vegetable gardens from red soil around the village to the sandy terrain to the south. Not only were the "moist", "soft" sands said to be more productive and easier to cultivate under declining rainfall, but a ruling by Hazohandatse elders on compensation for crop losses caused by stock trespass had accentuated the relative costs-benefits in favour of the sands. This ruling held that, while stock-owners would remain responsible for compensating damage their animals caused in garden plots on the sands, on the "redness" the onus would be on the farmer to protect his crops. To demarcate the two zones a hedge had been planted right along the escarpment. North of this line cattle and goats could be left to graze unherded on open grassland and in forest-park. South of this line vegetable plots on sands could be left unfenced. The upshot was that people became even more disinclined to garden on red soil.

Since then farmers had been clearing native bush from the hillocks and cactus hedging from between existing plots in order to expand their vegetable gardens on the sands. As a result, the sandy zone was being transformed from *bocage* into something analogous to the vast open fields (*valabe*) that French geographer René Battistini in the 1960s described for the Mahafale coastal plain. Battistini had seen these wind-blown *valabe* as especially prone to soil erosion, and recommended a remedial programme of *raketa* windbreaks (Battistini 1964:170–71). Some Hazohandatse commenting their own actions voiced similar fears. Up until then vegetable gardens on the sandy terrain had been considered permanent – lands which, informants insisted, would continue to crop well over many years without ever needing to be left fallow. De-shrubbing/de-enclosure also raised the spectre of increasing boundary disputes, leading some farmers to try to re-demarcate their individual plots in the vast open field by marking out the now invisible lines with stones.

Individual land holdings in Anjedava differed in size and quality. Some farmers, generally higher ranking villagers associated with senior descent lines (*ondate be raza,* 'people of high ancestry'), had inherited more land and better land than they could work with their disposable labour while other households had insufficient land to meet their subsistence needs. Of the latter those with limited access to gardens on sandy terrain were clearing forest from the red soil in order to access additional arable land.

One of the first steps a farmer took when creating a vegetable garden on fire fallow (*tetek'ala* lit. cutting in woods) was to plant a line of prickly pear (generally *raketa sonjo,* a variety of *Opuntia ficus-indica*). The high value of this particular variety as a human food and dry season fodder was a strong incentive for these plantings (Middleton 2009). Indeed, some farmers had been clearing forest specifically to expand their *raketa sonjo* holdings in the belief that the plantings would crop sooner on forest burns than on alternative free land. In many cases, though, the plantings were initially made as a quick and cheap way of staking a claim in (disputed) land.

Thus far, I have described how farmers were creating fire fallows in order to plant *raketa saonjo* in the woods around Anjedava for economic reasons. But the practice was also being encouraged by tenurial insecurity among those who "owned" more land than they could work. As one *mpizaka* (elder who mediates community disputes) explained, under Hazohandatse *fomba* (custom) rights in fallow (*moandrae*, "patrimonial fallows") are supposed to exist in perpetuity:

> Hazohandatse fallow (*moka*) has an owner (*tompo*). The field created by your ancestors/grandparents, *moka,* can't be seized by others, even when the trees grow very tall. Even when great wilderness, others can only beg permission to work your *moka*. If you go away, kin can work it in your absence but they have to give it back when you return. A hundred years [may pass] and *moka*'s not lost, so long as family is still around.[4]

In practice, he explained, tenurial security was more complicated because, in contrast to Hazohandatse custom, *vazaha* only recognise "ownership" in cleared and cultivated land. The term *vazaha* ("stranger", "foreigners", "white") here refers to Malagasy nationals, state officials, specifically the local *garde de forêt* who settled any land disputes that *zaka* (informal local councils) had been unable to resolve. As another villager commented:

> Nowadays it's not enough to say this is ancestral land because it's government land. The first thing they ask is: "Where are your crops?" Land no longer has an owner when the vegetation grows tall.[5]

*Vazaha* interpretations posed particular problems for entitlement in secondary forest on red soil. Rights in swidden-fallows were supposed to be permanent and inalienable yet it was implicit in the dynamic of swidden-fallowing that the period of active cultivation itself was always of limited duration. To retain "ownership" (usufruct) in the context of possible state intervention one needed to maintain soil in cultivation; yet the "redness" required "resting" from time to time. As a way of documenting "occupation", and thus of defending ancestral rights in swidden fallows, prickly pear had several advantages. It was cheaper and easier to establish on red soil than most other perennials (exotic or local); it was exempt from local taboos on sowing or planting fruit trees/bushes; and, most importantly, it was accepted as evidence of "occupation" by the state. Called to adjudicate a recent dispute over newly fired woodland on the boundary between Anjedava and the next village, the local *garde de forêt* had asked each

4 Toliandro, Anjedava, 15/5/1991.

5 Manjo, Anjedava, 14/6/1991.

party to show him the fruit trees he had planted on the plot. The man considered by most villagers to be its rightful owner (such had been the judgment of the *mpizaka* to whom the dispute had first been put) could only assert that the land was his "grandfather's fallow" whereas Malimana, the resident of Anjedava village who had appropriated the plot of land, was able to point to the *raketa* he had planted on it so the *garde de forêt*'s decision went his way.

Initially, there appeared to be an overlap between local and state positions (as the latter was interpreted for local contexts by the *garde de forêt*) because the elders also made recourse to planted tree stock as evidence of ownership (*famantara i tompo,* lit., proof of owner) when adjudicating land disputes. However, a closer look at their procedures revealed important differences in the types of evidence admitted. First, the *mpizaka* made reference to overgrown stands of *raketa* planted by people long ago, people who were now dead. This was because they upheld first clearer's rights. Secondly, the *mpizaka* drew on their extensive knowledge of the *lily*, a corpus of oral histories relating to the ancestors and the interclan accords that define access to natural resources (water, pasture, forests), in effect the unwritten rules or social arrangements that in Anjedava constitute customary tenurial law.[6] By contrast, the *garde de forêt* considered only more limited and time-shallow evidence. Basically, he looked for evidence of current occupation which he equated with recent planting activity. For the *mpizaka*, however, "ownership" was neither evidenced by nor reducible to *de facto* occupation.

The vulnerability of first clearer's rights in fallow – the cornerstone of the *mpizaka* tradition – before state interpretations was driving those villagers, whose existing vegetable gardens were more than adequate for their subsistence needs, to turn over tracts of forest to *raketa saonjo*

[6] I use the term "customary land tenure law" in this chapter to refer to ideas that Hazohandatse themselves hold about "ancestral custom" (*fombandrazañe*). In this context it is important to note that, although the concept of "customary law" is recognized and given importance in Madagascar state law, its actual content is not codified in national legislation. Blanc-Jourdan (1964) describes how, given the sheer diversity of practices that constitute local *fomba*, even between proximate villages (as this chapter shows), and the impossibility of codifying these for every region or sub-region of Madagascar, it was decided at the time Madagascar attained its formal independence to recognize the importance of "customary law" in state law but to delegate its definition in the particular instance to local traditional authority. As for *published* descriptions of so-called "Malagasy customary tenure", these, as I have noted, tend to be extrapolations from Merina, Betsileo, or Betsimisaraka ethnography (e.g. Olson 1984) and cannot be assumed to represent practice throughout the island.

simply in order to maintain visible long-term ownership in patrimonial fallows against land-short farmers who, as Toliandro put it, "laid claim to land to which they were not entitled and then appealed to the state when judgement went against them in village courts".[7] Citing Malimana's recent victory over the rightful owner, Toliandro concluded that to defend one's patrimony one needed to "domesticate land" (*miharehare tane*) by clearing and planting. "To leave land without standing crops, I'm afraid that's asking for trouble because this is now a foreign land."[8] Since, however, the "forest-park" indigenous to this region contains many fodder species (Poupon 1950) wealthier villagers faced a conflict in resource management between forest clearance in order to safeguard latent property rights for their descendants and retaining forest as vital browse and shade for their stock. Felling a little forest to plant prickly pear usefully resolved this dilemma because prickly pear was accepted by the state as evidence of occupation while also providing an invaluable supplement to forest browse, most notably succulence in times of drought.

I suggested that in settling land claims history mattered far more to the *mpizaka* than to the government official.[9] Villagers had been particularly disturbed by the *garde de forêt*'s recent decision because it allowed Malimana to claim a plot of sacred woodland. Known as "Blackforest" (*Alamainte*), with the local hardwood species that typically indicate "ancient wood" (*ala faharazañe* lit. woods of the ancestors' time), this sacred woodland had survived because of its association with the prayer posts of the Andrianjohoñe dynasty which ruled Karembola three centuries before. In 1991 most villagers were reluctant to enter the wood without first making a sacrifice and almost all still observed its "great taboos". Against communal sentiment, Malimana had secured "authorization" from Madagascar's Water and Forest Service to fell mature forest and desecrate taboo land.[10]

[7] Toliandro, Anjedava, 15/5/1991.

[8] Ibid.

[9] Although state law may encompass historical dimensions (Sandra Evers, personal communication, 2010), it seems unlikely that official tribunals would take account of the deep histories of entitlement in secondary forest that are recognized by many Hazohandatse farmers, especially as state law for over a century has made *effective occupation* (defined as clearance and cultivation) the primary criterion in successful claims in untitled land. Recent reforms retain this principle (Government of Madagascar 2005).

[10] It is important to note that when I refer to state arbitration in the Karembola context I mean arbitration by forestry officials. It is far from easy for rural Malagasy to refer a tenure dispute to the land tribunal when dissatisfied with the outcome of customary process, and indeed few Karembola do. In theory the forestry service was responsible for granting

What exactly had been "authorized" by the *garde de forêt*, and under which piece of legislation, was not entirely clear. Villagers themselves stated that Malimana had "purchased" (*nivily*) ten hectares of Blackforest from the *garde de forêt* for the price of two thousand Malagasy francs per hectare and two goats. "Purchase" here was a euphemism because everybody, including Malimana, knew that the transaction would never be registered at Fort Dauphin Land Registry and the state therefore remained "owner of the wood" (*tomponala*). Yet by the transaction Malimana had clearly acquired some entitlement/dispensation which carried weight within the community, even if it fell short of registered title. Many of the disputes arbitrated by the forestry official appeared to be settled in a kind of no-man's-land between "state" and "customary" law, generating a kind of sub-prime access to land.

It is tempting to contrast the ill-informed, possibly corrupt, practice of the *garde de forêt* (a man of Bara ethnicity and thus an outsider to the Karembola region) with the detailed landscape/kinship histories that the *mpizaka* deployed in their deliberations, and to attribute the loss of ancient forest to the bureaucrat's failure to understand the complex nuances of local systems of land management and the meaningful social relationships that exist between people past and present in respect of the land. To idealize local dispute mediation processes (*zaka*) would be wrong, however, when for several reasons they were weighted against the land poor. First, the *lily* – the histories whose interpretation broadly shaped everyday patterns of resource management in the Anjedava region – were by their nature weighted to upholding the rights of "senior" descent lines over those of "junior" or "dependent" lines, especially those whom Hazohandatse euphemistically term *mpiasa* ("workers" i.e., the descendants of former slaves and uxorilocal men). Secondly, knowledge of the *lily* was itself unevenly distributed among villagers. Thirdly, the elders who mediated land disputes were generally drawn from senior lines. In such circumstances, tenurial disputes were always more likely to be settled by traditional authority in favour of the status quo.

I have thus far highlighted contrasts between the positions taken by the forestry official and the *mpizaka*; but it is equally important to note

permits to exploit forest resources while land title was a matter for the *Domaine*. But it may not be uncommon in rural Madagascar for forest guards to regulate tenurial disputes, at least in districts such as Karembola where there are few incomers with the money, knowledge, and motive to pursue formal land claims. The incident recounted here predates the policy of *reserves foncières pour le reboisement* (RFR-2000-383) which therefore cannot be used to interpret the forestry official's ruling on the Malimana case.

that villagers also differed amongst themselves in the importance they attached to "history" in their understandings of how the practice of swidden fallow (what elsewhere in Madagascar might be called *tavy*) related to ownership rights over time. Almost all informants described the work involved in creating a field from woodland as a process of self-realization, an act of value creation, the manifest mark of a "robust person who accomplishes things" (*mahaozatse, mahavita raha*). However, in the opinion of those who had inherited extensive land it was the original act of clearing that was definitive of property rights. It was the work undertaken by their ancestors that had created ownership (and thus unequal land holdings) for all time. By contrast, the land poor were more inclined to see secondary forest around Anjedava as a common resource upon which any member of the community was free to draw. Because they saw forest clearance as an ongoing process that remade ownership in the present, they placed a greater emphasis on *personal* effort invested in land in the here-and-now. As Malimana, the man who successfully laid claim to someone else's fallow, put it:

> ancestral forest doesn't belong to anyone [in particular]. It's land you just create, which subsequently reverts [to forest]; it's not like a field which is inherited when a person dies.[11]

From this perspective, it could be argued that state interference – albeit by a dysfunctional forestry service poorly informed of the regulations and not adverse to taking bribes (cf. Montagne and Bertrand 2006) – was at least helping to redistribute land from land-rich households by bestowing *post hoc* "state" approval on the *de facto* seizure of forest fallows by the land poor. However, against such a position, it might be contended that customary tenure had always made provision for the borrowing and lending of land, and that accessing land by means of state-sanctioned seizure had only become necessary because those with excess land were no longer willing to allow others to cultivate it temporarily for fear of losing their hereditary rights. We might even be tempted to idealize customary practice (as it supposedly existed prior to forestry service and other state interventions) as a highly flexible mechanism of land allocation that adjusted garden plots to ongoing socio-demographic processes, including the inevitable but essentially unpredictable contraction and expansion of individual households over their lifecycles. In fact, there is no way of glossing the often asymmetrical connotations of the historical practice of land

---

[11] Malimana, Anjedava, 16/4/1991.

loans and land allocation in Karembola. While much land was indeed loaned and borrowed between kin, that is, essentially between equals, the ability to lend (poorer quality) land was also a key way of attracting and incorporating outsiders as clients and dependants. In these instances, cultural claims to authority, to ownership of land, and to possess the spiritual power to dispense fertility and blessing to people, land, crops and livestock went hand in hand.

For this reason many villagers saw men who referred land claims to the state for arbitration as challenging beliefs and entitlements that underpinned traditional authority. In this respect, we can say that *raketa* planting in woods was being driven by two contrary political tendencies within Hazohandatse village communities: the one questioning the hierarchical idioms that were supposed to order relations between people in respect of land; the other seeking to uphold them.

It is important to note that, under even under élite interpretations of Hazohandatse customary tenure, rights in secondary forest are not exclusively vested in an individual. While only the descendants of the farmer who first cleared a plot are said to be entitled to clear it again for cultivation, others are permitted to use the fallow temporarily, to graze cattle, for instance, or to collect wild foods (plants, birds, small mammals). In other words, even in élite readings of the landscape, there was a distinction between the land itself and the secondary growth. In this sense even the land wealthy accepted that the more secondary growth takes over, the more the land becomes a community resource. However, trees or shrubs such as prickly pear that were planted by the individual, that is, that grew through human agency, are supposed to remain the property of the person who planted them (or his or her descendants), however long the land is left fallow. From this perspective, the *raketa* hedging stands among the secondary vegetation as a "kind of postcard to future generations", an enduring "sign of (individual) ownership" in (temporarily) common land.[12]

The problem with a living fence as over paper title, however, is that *raketa* can turn into what Karembola term *ala* (forest), and its significance for land rights can then only be deciphered with reference to history. There inevitably comes a point when the power of prickly pear stands (or indeed any tree) to document specific lines of ownership depends on memory of past relationships, and with memory comes room for interpretation and contestation, and the original planter may be forgotten or denied. This process is accentuated in the case of invasive varieties.

---

[12] The postcard analogy is borrowed from Gow (1995, 56) who however criticises such semiotic interpretations.

It might be appropriate to pause here for a moment to underscore two points. First, while local people themselves often invoked sharp contrasts between *vazaha* concepts of land ownership and Hazohandatse "custom", these dichotomous representations, so typical of indigenous Malagasy discourse on land tenure, are not necessarily to be taken as accurate representations of historical relationships between the local and the national. Historically, as the data discussed here show, state law and local practice in the Karembola region have continually shaped and reshaped each other in an ongoing co-evolutionary process. At the same time, and partly because of the way this co-evolution has impacted on the constitution of local practice, the characterisations and interpretations of tenurial "custom" that were offered by local people were often diverse. Moreover, historical interactions between state law and local practice had not only created multiple, contested interpretations of "ancestral custom" upon which villagers selectively drew in order to substantiate their particular claims. But state forestry officials, called to judge land disputes that village communities had been unable to resolve, themselves also drew on a hybrid confection of local and state rules.

### *Proto-Paddocks*

The data discussed thus far have indicated that practices of land use and land tenure in and around Anjedava were complex and evolving. And from this perspective, one of the most interesting developments was the appearance of what Hazohandatse termed *goloboñe,* and I shall call proto-paddocks,[13] in the zone immediately around the village where, instead of allowing former fields to lie fallow, many farmers had been re-enclosing their land. The aim may have been to intensify livestock production by creating "private" grazing land for improvement as distinct from common access rangeland over whose condition the individual herder has little or no control. Some *goloboñe* were planted with forage *raketa,* and there were signs that prickly pear had been recently burned from many more to feed cattle during the prolonged drought. Otherwise I could detect little conscious attempt at improvement, by seeding grasses for instance, and indeed

[13] As former fields on land *between* the village settlement and wider *terroir, golobo* were both far larger and differently located than the traditional kraals or cattle pens that are found *within* the village, where cattle, especially milch cows, are stockaded by night. In terms of size, locality, and function, the term *golòboñe* in Anjedava usage seems to correspond more closely to the definition offered by Sarah Fee and Narivelo Rajaonarimanana (1996) for the term *kolòbo* or *kolòboñe* in the Tandroy dialect ('espace (au village ou dans la forêt) entouré de haies pour parquer boeufs et chèvres') than to the gloss they provide for *golòbo* n. *golobon-tana:* 'espace dans un village entouré d'une haie de sisal ou de *raketa*'.

throughout the nine months I lived in Anjedava (January-September 1991) the proto-paddocks stood mostly empty, village cattle being left day and night to browse in woods. It did occur to me that these empty spaces might be as much about keeping stock out as keeping it in, and that the long-term aim was to allow forest-park to regenerate in old gardens around the village where large numbers of goats otherwise allowed only a few unpalatable plant species to survive. No one articulated such a proposition to me, however, and overall I suspect that this embryonic enclosure movement was being driven by the same concerns around tenurial security that were driving Hazohandatse to cut prematurely in the forest if only to plant prickly pears. As one villager noted cryptically, "*goloboñe* are better than *moka* because they still have fences, still have owners".[14] In other words, hedging (enclosure) was critical to the definition of *goloboñe*. Failure to lay or to maintain the hedge meant that a former field/patrimonial fallow became indistinguishable from *moka* or common cattle grazing (*tane añombe*) that "belongs to everyone". In the past the temporary rights of villagers to graze cattle on self-seeded or resprouted vegetation would have been compatible with the enduring rights of a particular household in the land as indicated by planted trees. However, state insistence upon linking "ownership" with occupation has made the practice of multiple rights in fallow risky. In this respect, the functional indeterminacy of the proto-paddocks was indicative of a land use system in transition where the outcomes of farming decisions were generally uncertain for all sorts of reasons (including possible state interference or climate change) and where farmers accordingly were hedging their bets.

From this overview, it will be seen that land rights in Anjedava had a complexity and a fluidity which cannot be understood in terms of simple, dualistic contrasts between, on the one hand, a "modern" system of formal land registration based on private property and, on the other hand, "customary" or "traditional" tenure practices based on communal land rights. Not only were different types of land (gardens, pasture, secondary forest,...) subject to distinct tenure arrangements but individuals often had overlapping but unequal rights in these different types of land. This complexity and flexibility reflected the basic truth that successful dryland farming is based on the skilful exploitation of multiple microenvironments, and the more numerous and diverse the econiches a household can access, the better equipped it is to manage climate variability.

---

14 Masiñeavy, Anjedava, 1/7/1991.

At first sight the data collected in Anjedava appear to contradict the most common argument for land titling in development literature, viz. that farmers will only invest in perennial crops and soil conservation measures when provided with security of tenure (World Bank 1990: 5). In Anjedava, the contrary seemed to be happening: insecure land rights were stimulating one, possibly two types of intensification: fruit bush planting and proto-paddocks. In truth, the matter was more complex because tenurial insecurity was also pushing some farmers to clear more native forest than they needed in order to plant the perennials that state officials require as evidence of ownership.[15] In effect, insecure land rights (perceived or actual) were driving contradictory impulses: reforestation *and* deforestation, agricultural intensification *and* biodiversity loss.

### *Land Rights on the Periphery of a Reserve*

I now turn to land rights issues on the periphery of Cap Sainte Marie Special Reserve (*Réserve Spéciale du Cap Sainte Marie*). This reserve is located some 65 km east of Anjedava, at the southernmost tip of Madagascar, where the headland dramatically falls away to the sea (see Map 1). Decreed a "Protected Area" in 1962 (that is, around the time that tree-planting was made a statutory obligation in Madagascar), this 1750 hectare reserve was more or less neglected until the 1990s when the rise of the conservationist lobby in Madagascar coupled with growing global interest in "spiny dry forest" led to stricter enforcement of controls (Nicoll and Langrand 1989: 123–5). By 2002, when the data reported here were collected, draconian sanctions (fines and prison sentences) against villagers had eliminated all garden plots and cattle-pens from inside the reserve, and reduced the number of goat-pens within its perimeters to two. The aim in comparing the Anjedava and Cap Sainte Marie cases is to show how a rather different set of circumstances produces strikingly different configurations of the relationship between local land tenure, external interventions, and exotic plants.

At the time of fieldwork stricter enforcement of the legislation relating to the reserve was impacting negatively on resource use and food security among the local population (c. 2.682), which, apart from one fishing

---

[15] It is not uncommon, in former African colonies where land is largely unregistered, for a policy that holds land title only exists in respect of cleared land to promote deforestation (Mortimore 1998, 11; World Bank 1990, 104).

village inside the reserve, was located in the so-called *zone périphérique* to the north of the reserve. Villagers considered secondary forest within the reserve to be part of their swidden-fallow, in effect their clan's oldest cultivable land. The loss of traditional grazing rights was particularly keenly felt by stock-owners because the best browse for cattle, notably *ametse* (*Euphorbia stenoclada*), was found inside the reserve. Moreover, forest park to the north and the east of the villages belonged to other clans while the quality of traditional grazing lands to the west was said to be poor. Officers from ANGAP (*Association Nationale pour la Gestion des Aires Protegées*/ National Association for the Management of Protected Areas), the organisation charged with managing the reserve, had been monitoring the impacts of stricter enforcement of exclusion on local livestock production, and even they reported that stock densities in the "peripheral zone" had increased substantially while average per capita cattle holdings had declined.

The impacts of land reservation on local cropping systems were less immediately obvious at the time of fieldwork partly because, according to informants I interviewed, at the time the land was officially surveyed in the late 1950s in preparation for the legislation that created the reserve, local swidden-fallow cycles had been at the stage where villagers' vegetable gardens were mostly located north of the territory that was appropriated for the reserve. In other words, few villagers had been displaced from actively cultivated land. Indeed, it was noticeable that although most informants in 2002/2003 described land supply in the Cap Sainte Marie region as "tight" when compared with Karembola districts further west, many attributed this more to comparatively low local rates of labour outmigration and to local tenure practices than to the loss of fallow and secondary forest which had been withdrawn from local agricultural systems and allocated to the reserve. For instance, they noted that, in contrast to Anjedava where the observation of male primogeniture privileged the eldest son over younger brothers and female siblings, parents at Cap Sainte Marie tended to divide their fields between their (often numerous) sons and unmarried daughters on a more equal basis. Over time, individual landholdings had become fragmented and cultivated land as a proportion of available land had risen. As a result, they argued, many families had insufficient land of sufficient quality to meet their subsistence needs. Only some then added that in the past the land poor would have cut a field in secondary forest/fallow to the south of the villages but that this option had ceased to exist with stricter patrolling of the reserve.

In theory, trapping and harvesting natural products within the reserve were also prohibited (and had been so since 1962); but the impact of these particular restrictions on local subsistence and survival strategies was difficult to ascertain because villagers still enjoyed rights of access through the reserve. Overall, though, there can be no doubt that the reserve had become a direct competitor for scarce local resources while offering little in the way of income or alternative employment opportunities to compensate. On average 350 tourists visited annually, and political troubles in Madagascar meant few had come in 2002. The reserve had conspicuously failed to deliver on the conservationist promise to ensure that livelihoods on the peripheries of reserves would not suffer from biodiversity protection objectives (McNeely and Miller 1984; Wells and Brandon 1992; Wright 1997).

The curious thing, however, was how little villagers at the time of fieldwork complained about the reserve and its impact on land "rights" and resource use. On the whole, they were preoccupied with another subject: invasion of their fields and grazing lands by a spiny opuntia, known locally as *raketamena* ("red prickly pear") on account of the colour of its fruit. A relatively recent arrival in southern Madagascar (probably dating to the 1970s), the plant had been propagated by villagers in the days before the dangers were known. By the time of fieldwork it was said by most villagers to be impossible for unmechanised farmers to eradicate. Two WWF-commissioned reports had stressed the dangers the infestations were posing to the "survival" of the local human population and to biodiversity within the reserve, and claimed that an initiative for eradication/control had "local community" support (ANGAP et al. n.d. [2001]; Randriamampianina et al. 2002).

A campaign specifically against *raketamena* (rather than, say, against all introduced Opuntia in southern Madagascar, many of which are very useful to dryland communities) was attractive to the WWF. This was because there *appeared* to be a coincidence between its own biodiversity conservation objectives and local farmer interests. It is true that, as a signatory to the 1992 Convention on Biological Diversity, Madagascar had agreed under Article 8(h) "to prevent the introduction of, control or eradicate those alien species which threaten ecosystems, habitats or species" (cf. Cronk and Fuller 1995; Mooney and Hobbs 2000; Sandlund et al. 1999). But since the 1990s, when the rhetoric of *integrated* conservation and development became *de rigueur* for donor agencies active in Madagascar, it has become important for the WWF to argue the case for any projects or interventions on both ecological *and* socio-economic grounds. A programme of *raketamena* control or eradication seemed to WWF staff to qualify on both counts.

In truth, however, local support for an initiative against *raketamena* was far from unanimous. Certainly, of the 100 plus villagers I interviewed in the Cap Sainte Marie region, none denied that *raketamena* was a prolific spreader that was posing serious problems for local farmers. However, opinion was deeply divided over what should happen to the plant. Some villagers called for its swift eradication. But others protested vehemently that the fruit was keeping them alive. In effect, the plant had significantly different cost-benefits for individual community members, corresponding largely to variable patterns of livelihood. For many resource-short households struggling to survive the 2002/2003 drought with limited or zero livestock and insufficient, often poor quality, arable land, the value of *raketamena* as a famine food outweighed its costs (Middleton 2012, n.d.). As one farmer, responsible for fourteen children including several orphans, put it:

> The rich hate *raketamena* because it kills the land. But I value it because I'm poor [*ondate tsymanandraha* lit. "people who have nothing"]. The wealthy ... those with cattle, with chickens, are willing to give up [*raketamena*] because they [have things they] can sell at market [to buy in food]. I've no cows, not a single one. How can we get money? There's no [paid] work here. It's dead. We just cultivate our fields for better or for worse. We go to work on *raketamena*, we go to sleep on *raketamena*. It's our livelihood right now. We're in famine times.[16]

Wealthy villagers took another view. Not only had they no need for *raketamena* as a food but they blamed declining livestock numbers on the plant. Glochids (tiny hairs or spicules on the fruit) were said to collect in the animals' digestive tracts while consumption of the leaf pads allegedly caused a fatal bloat. *Raketamena* invasion was not therefore perceived as a cheap, fortuitous means of improving degraded rangeland. On the contrary, according to those who owned cattle, sheep or goats, the plant was aggravating the problems of feed supply by invading bushy clearings and forage *raketa* plantations. Ultimately, *raketamena* was undermining the material bases of local élite power, which depends partly on achieving the above average levels of crop and livestock production that are required in

[16] Befañahe, Befela, 10/1/2003.

order to be able to host the feasts and maintain the networks of ceremonial exchange that confer influence and prestige in the Karembola region.

As at Anjedava these were internally divided communities in which marked wealth differentials were associated with contested patterns of resource management, access and use. However, the dynamics of conflict were configured differently. Anjedava had a relatively low cultivated fraction, and the supply of forested land was still sufficient to allow some choice in locating fields. Here prickly pears had been embroiled in disputes largely between individuals over specific plots of land. A perceived growing scarcity of land was accelerating its acquisition and "privatisation" by farming households, anxious not to be left with insufficient land when the supply was finally exhausted. But the basic problem in Anjedava was *unequal* distribution of land at the household level rather than absolute supply of land.

By contrast at Cap Sainte Marie, virtually all cultivable land outside the reserve was in use and/or claimed by someone. Ancestral fallow had been appropriated for biodiversity conservation, and accordingly the practice of planting *raketa* to stake claims in forested land had more or less ceased. Farmers were no longer able to rely on the practice of fallows to replenish the soil while falling cattle numbers meant that systematic manuring of arable lands had if anything decreased.

Curiously, however, informants seemed more exercised about land rights at Anjedava where land was still available than at Cap Sainte Marie where land was perceived as "tight". We might employ the distinction David Stanfield (1995: 1) has drawn between objective and subjective security to capture the contrast. In Anjedava farmers *felt* at great risk of losing their land reserves to other farmers in the context of possible state interventions whereas at Cap Sainte Marie land reserves had already been appropriated by a more powerful state/biodiversity initiative. (In 2003 villagers seemed unaware of conservationist proposals to increase still further the size of the reserve [Rakotomalala 2008: 173, 176].)

In both instances, local farmers lacked formal title but in Anjedava cactus hedging was still a component of an active dynamic of claim and counterclaim whereas in Cap Sainte Marie cactus hedging, still visible inside the reserve, remained only as testimony to the former use of land by local people (cropping, stockpens) but had ceased to count from the perspective of current land rights. Here the prickly pear hedging figured in narrative only to indicate a time before the land was appropriated by *vazaha* (or rather before *vazaha* began actively to enforce an appropriation formally effected in legislation thirty years before).

On the periphery of the reserve, the transmutation of *raketamena* from cultivar into weed was threatening not only economic productivity but also enforceable social landscape conventions. The use of prickly pear to document ownership ("occupation"), under both state and local tenurial rules, had been based on cultivars, varieties that could be relied upon to stay strictly in their place. Even when *raketa saonjo* grew into thickets, or seemingly ran wild in forested places, the *mpizaka* could safely make the assumption that someone, once, had planted it. By contrast, *raketamena* plants were self-seeding, capable of popping up all over the place, undoing social acts of planting and hedging, confusing private and common land. Even where it had been planted to begin with, villagers were now often too ashamed to admit ownership. There was consequently a pending crisis in reading the landscape as "owned", "bounded", and socially constituted space. Fields, grazing lands, forest-park, and sacred copses, that is, distinctive social habitats whose particular meaning and use were delineated by cultural conventions, were rapidly losing their separate identities as they all became land where "nothing but *raketamena*" grew.

In both communities rich and poor were fighting over "ownership of the land, but the meaning of that "ownership" and the politics of community had undergone subtle shifts. In both communities "land rights" were at issue but in Anjedava the battle was between individuals or families over control of and access to specific plots of land whereas in Cap Sainte Marie it was a generalised dispute between factions (mainly poor and rich) over appropriate land use on a more regional scale. There was a degree of crossover: land use was an issue in Anjedava (as, for example, when stockowners wanted to retain local forest as browse whereas land-short households with few or no cattle were more likely to view it as potential arable land) while land scarcity clearly underlay the disputes over *raketamena* at Cap Sainte Marie. But in terms of how ongoing battles over natural resource management were phrased rhetorically and ideologically by informants in the Cap Sainte Marie region, social conflict over an invasive plant now largely masked disputes between villagers over unequal access to scarce land.

The political field had also been reconfigured. Whereas in 1990s Anjedava the continued dominance of "big men" had been under challenge from villagers, such as the wily Malimana, who refused to accept their judgement and went "forum shopping" to the state, by 2002 the influx of unprecedented numbers of development and environmental agencies into the Karembola region, in the wake of the 1992 famine and with the rise of global governance in Madagascar more generally, had led to a retrenchment of élite power.[17] By exploiting a multiple clientelism with the WWF and the World Food Programme and their various satellite organisations, "community leaders", at Cap Sainte Marie especially but to a lesser extent throughout Karembola, had been able to consolidate their positions, reassert control over natural resources, re-establish patronage networks, and protect and even advance their own land use and resource management needs (Middleton 2012, n.d.; cf. Brockington 2006; Duffy 2006).

### *Broader Relevance of Case Studies*

It is perhaps inevitable that a modern state searches for uniformity and simplification through the imposition of consistent legal codes (Scott 1998; World Bank 1990: 8). However, the substitution of universal principles based on exclusive title for multiple subsidiary user rights derived from local custom often deprive poorer households of the means to supplement insufficient cultivated food with a wide range of trapped and gathered foods and to derive some cash income by harvesting wild or semi-wild Malagasy periwinkle and castor oil seed for sale. Global experience suggests that graziers and gatherers in particular are seldom compensated for their loss of usufruct under land titling programmes (Cederlöf 2008: 120–21, 135; Shepherd 1991; Dekker 2003). From this perspective, it could be argued that the current "legal and institutional pluralism" that Pronk and Evers (2007) identify as the source of multiple social and ecological problems in East Coast Madagascar at least provides a sphere for competing judicial principles to be regionally negotiated with respect to particular and overlapping rights in land.

A decline in cattle numbers would be welcomed by many conservationists who associate stock-keeping with overgrazing and biodiversity loss (e.g., Soarimalala and Raherilalao 2008: 159). However, it is important to remember that indigenous "forest-park" partly owes its survival to its value to stock-owners (wealthier community members) as browse and shade. The land poor would sooner turn much of it into arable land. Fewer cattle might well reduce local need for native forest, and lessen

[17] On global governance in Madagascar see Duffy 2006 and more generally Diehl 2001; Harrison 2004; Murphy 2000; Paolina et al. 1998; Young 1997. I use the term "forum shopping" of those who appeal to the forest guard for arbitration, rather than in its usual sense of referring a tenure dispute to the land tribunal. On "forum shopping" see Benda Beckman 1981.

local motivation for forest conservation. Global literature indicates that production shifts from cattle breeders to increasing numbers of small farms has caused partial deforestation in other parts of the world (Cederlöf 2008: 128; Shepherd 1991).

Nor may the reduction of cattle numbers necessarily have beneficial effects for local livelihoods. Given the inherent variability of the Karembola climate, it is difficult to see how small producing units would survive without the integration of crops and livestock. Specialised production is more dependent on external inputs which tend to be uneconomic in dryland conditions (Mortimore 1998: 95). Yet until and unless water supplies are greatly improved, successful pastoralism in Karembola will continue to depend on herd mobility and flexible access to land, i.e. on the present complex body of rights that is at odds with the land codification being proposed (cf. Scoones 1995; Leach and Mearns 1996: 29).

The data presented in this chapter suggest that the relationship between tenure incentives, conservation and sustainable development is a complex issue, and that land registration may not be the panacea for Madagascar's "problems" that many policy makers and shapers assume. It is difficult to see how land titling by itself might address the myriad problems touched on here: biodiversity loss, environmental degradation, food insecurity, sustainable livelihoods, invasive species, social inequality, let alone reconcile them. Some of the evidence from Anjedava appeared to support the case for land titling but on closer inspection proved more complex because a number of converging pressures was incentivising forest clearance and *raketa* planting, not tenurial insecurity alone. These included the need to expand food production in response to a perceived increase in the frequency of droughts. Moreover, farmer behaviour to date predicts that the prospect of a land registration programme that was expected to accede title to those who can claim effective occupation (Government of Madagascar 2005) would simply stimulate further deforestation and *raketa* planting as local people rush to stake individual claims in all available forested land (see Lund 1993 for a cautionary tale from Niger). Even supposing that land titling was followed by on-farm intensification (which seems unlikely because of the capital costs and the risks of such investment under dryland conditions) much endemic biodiversity would have been lost for good. The well-rehearsed argument about a tree-planting revolution taking place wherever farmland expands and becomes more permanent under individual tenure (Mortimore 1998: 32) glosses the critical distinction between exotic and native species.

Most hardwoods indigenous to Karembola are notoriously difficult to propagate artificially and regenerate only within the forest.

With respect to Cap Sainte Marie Special Reserve, "encroachment" has been curtailed by stricter enforcement of existing regulations rather than by land titling. The application of prison sentences and punitive fines during the nineties reduced farmers' economic incentives to ignore the prohibitions applying to the reserve. (As for whether the increasing exclusion of villagers from the reserve will actually protect biodiversity within it, this is debatable. The answer depends partly on the extent to which reserve biodiversity has been shaped and produced by centuries of human management and use [cf. Balée 1989; Fairhead and Leach 1996].)

Would an alternative tenurial system have ensured better outcomes for villagers on the periphery of the reserve? It is difficult to see how a programme of land titling alone would compensate villagers for loss of access to resources in the reserve (including the loss of the latent right to resume cultivation of ancestral fallows sometime in the future) or enable individual farmers to make better use of their (now much reduced) land. As for alleviating poverty and improving food security, the narrative of land titling ignores the fact that agricultural intensification requires capital investment, and that for most villagers local opportunities for income diversification are limited. Those with capital to spare tend to invest it in cattle and tombs (social relations) rather than in capitalizing agriculture because of the high risks of crop failure under prevailing environmental conditions in the Malagasy Deep South.

Overall the data reported in this chapter provide little basis for sharing Pronk and Evers' hope that the *Programme National Foncier* [PNF] initiated by GoM and World Bank will give rural Malagasy the security of land access that legal and institutional pluralism has been unable to provide (Pronk and Evers 2007). Not only are there very serious questions about the process by which rights in land will be settled (cf Bertrand 2008) but advocates of PNF fail to explain how a programme of individual land titling will address the needs of the land-poor. While insecure access to land may be "one of the main determinants of poverty" (UNDP 2000: 32), it is less easy to see how codification *alone* (that is, without either an element of significant redistribution of land and/or further deforestation, not to mention external funding for capital investment in land) will resolve what appears to be the key problem in many areas of Madagascar: scarcity of fertile land under rapid population growth (estimated at 2.6 per cent/year in 2006 (World Bank 2006b).

Nor do the data offer unequivocal support for those who advocate common property solutions to biodiversity conservation and economic development. The problem with community tenure practices is that "communities" seldom are homogeneous, and access to and use of resources are often both sharply differentiated and contested between groups (cf. Graziana and Burnham 2002; Li 2002; Peluso 1992; Sundar 2000; Schroeder and Suryanata 1996). Herding and farming interests often put contrasting pressures on woodland resources under communal property management in semi-arid areas of Africa (Shepherd 1991). While group tenure may provide a satisfactory alternative to individual tenure in some cultures, the very fluidity of resource access under communal tenure may allow better-resourced community members to extend their resource base at the expense of the less advantaged (Mortimore 1998; Dekker 2003). Arguments for "community forests" in particular overlook the intricacies of existing land tenure practices and the diverging livelihood interests of poor and rich, and the fact that under swidden fallowing cycles (collective) woodland is always potentially (individualised) farmland and therefore its management cannot be separated from broader patterns of land use (Shepherd 1991).[18] Questions also exist as to whether communal tenure is adequately secure to stop governments from the arbitrary alienation of land.

Rather, the case studies presented here suggest that when questions around land tenure, environment and development in Madagascar are reframed to take account of ecological and socio-political specificities, it will be seen that there is not one single bundle of problems in the island but rather many different problems which take different forms for different people in different places and therefore that the range of potential solutions may be equally diverse (cf. Leach and Mearns 1996: 2–3). The more complex and site-specific existing tenurial arrangements are, the less appropriate the use of "blue-print" development strategies becomes.

[18] Recent studies of community forests in the spiny dry forest region of Madagascar have focused on southern Androy, a landscape of permanent fields and sacred copses (Elmqvist et al. 2007; Bodin et al. 200l; Tengo et al. 2007). These studies tell us little about forest governance and resource tenure under the swidden-fallowing systems that prevail in other parts of southern Madagascar. Tengo et al. 2007 include one case study from Karembola (Lavanono) but the data are poorly developed. They do, however, reveal a higher proportion of "private" forest in this locality than in southern Androy, a characteristic which is clearly in keeping with the tenure patterns for Anjedava I describe.

## *Bibliography*

ANGAP-FOFIFA-WWF (n.d.) Rapport de Mission à Cap Ste Marie (du 03 au 06 mai 2001) et Propositions des Actions Futures. Unpublished Report.

Balée, W. 1989. The Culture of Amazonian Forests. *Advances in Economic Botany* 7:1–21.

Barbera, G., P. Inglese & E. Pimienta-Barrios (eds). 1995. *Agro-Ecology, Cultivation and Uses of Cactus Pear*. Rome: FAO.

Battistini, R. 1964. *Géographie humaine de la plaine cotière Mahafaly*. Toulouse: Éditions Cujas.

Beinart, W. and K. Middleton. 2004. Plant transfers in historical perspective: a review article. *Environment and History* 10, 3–29.

Benda Beckman, K. von. 1981. Forum Shopping and Shopping Forums: Dispute Processing in a Minangkabau Village in West Sumatra. *Journal of Legal Pluralism* 19, 117–59.

Bertrand, A. 2004. The spread of the Merina people in Madagascar and natural forests and eucalyptus stand dynamics. In *Beyond Tropical Deforestation*, ed. D. Babin, 151–56. Paris: UNESCO/CIRAD.

Blanc-Jourdan, X. 1964. Development of a new law code in Madagascar. *Africa Today* 11(7), 7–9.

Bodin, Ö, Tengö, M, Norman, A., Lundberg, J. and Elmqvist, T. 2006. The value of small size: loss of forest patches and threshold effects on ecosystem services in southern Madagascar. *Ecological Applications* 16, 440–51.

Brockington, D. 2006. The politics and ethnography of environmentalisms in Tanzania. *African Affairs* 105 (418), 97–116.

Cederlöf, G. 2008. *Landscapes and the Law. Environmental Politics, Regional Histories, and Contests over Nature*. New Delhi: Permanent Black.

Clark, C., S. Garrod and M. Pearson. 1998. Landscape archaeology and remote sensing in southern Madagascar. *International Journal of Remote Sensing* 19 (8), 1461–77.

Cronk, Q. and J. Fuller. 1995. *Plant Invaders: The Threat to Natural Ecosystems*. London: Chapman and Hall.

Decary, R. 1930. *L'Androy (Extrême Sud de Madagascar)*. Paris: Société d'éditions géographiques, maritimes et coloniales.

Defoort, E. 1913. L'Androy. *Bulletin économique de Madagascar*, 2, 127–246.

Dekker H. 2003. *The Invisible Line. Land Reform, Land Tenure Security and Land Registration*. Aldershot: Ashgate.

Diehl, P. ed. 2001. *The Politics of Global Governance: International Organizations in a Changing World*. Boulder, CO: Lynne Rienner.

Duffy, R. 2006. NGO and Governance States. *Environmental Politics* 15(5): 731–49.

DuPuy, D. and J. Moat. 1996. A refined classification of the primary vegetation of Madagascar based on the underlying geology. In *Proceedings of the International Symposium on Biogeography of Madagascar*, ed. W. Lourenco, 205–18. Paris: ORSTOM.

Elmqvist, T., M. Pyykönen, M. Tengö, F. Rakotandrasoa, E. Rabakonandrianina, and C. Radimilahy. 2007. Patterns of Loss and Regeneration of Tropical Dry Forest in Madagascar. PLoS ONE 2(5).e402. doi: 10.1371/journal.pone.0000402.

Fairhead, J. and Leach, M. 1996. *Misreading the African Landscape. Society and Ecology in a Forest-Savanna Mosaic*. Cambridge: Cambridge University Press.

Fee, S. and N. Rajaonarimanana. 1996. *Dictionnaire Malgache dialectal – Français. Dialecte Tandroy*. Paris: Langues & Mondes/L'Asiathèque.

Fortmann, L. and J. Bruce eds. 1988. *Whose Trees? Proprietary Dimensions of Forestry*. Boulder: Westview Press.

Gade, D. and A. Perkins-Belgram. 1986. Woodfuels, reforestation, and ecodevelopment in highland Madagascar. *GeoJournal* 12(4), 365–74.

Government of Madagascar. 2005. Lettre de Politique Foncière. Texte officiel validé 3 mai 2005, http://www.cirad.mg/fr/foncier.php (accessed 13/06/2010).

Gow, P. 1995. Land, People, and Paper in Western Amazonia. In *The Anthropology of Landscape. Perspectives on Place and Space* eds. E. Hirsch and M. O'Hanlon, 43–62. Oxford: Clarendon Press.

Graziana, M. and P. Burnham. 2002. Legal pluralism in the Rain Forests of Southeastern Cameroon. In *Plural Resources and Local Livelihoods in Africa*, ed. K. Homewood, 177–97. Oxford: James Currey.

Harrison, G. 2004. *The World Bank and Africa: The Construction of Governance States.* London: Routledge.

Healy, T. and R. Ratsimbarison. 1998. Historical influences and the role of traditional land rights in Madagascar: legality versus legitimacy. University of Cape Town Proceedings of the International Conference on Land Tenure in the Developing World with a Focus on Southern Africa 27–29 January 1998, 232–42.

Jacoby, H. and Minten, B. 2007. Is Land Titling in Sub-Saharan Africa Cost-Effective? Evidence from Madagascar. *The World Bank Economic Review* 21(3), 461–85.

Kull, C. 1996. The Evolution of Conservation Efforts in Madagascar. *International Environmental Affairs* 8(1), 50–86.

Leach, M. and R. Mearns. 1996. Environmental Change and Policy. Challenging Received Wisdom in Africa. In *The Lie of the Land. Challenging Received Wisdom on the African Environment*, ed. Leach, M. and R. Mearns, pp. 1–33. Oxford: IAI/James Currey.

Le Houérou, H. 1996. The role of cacti (*Opuntia* spp.) in erosion control, land reclamation, rehabilitation and agricultural development in the Mediterranean Basin. *Journal of Arid Environments* 33: 135–59.

Li, T. 2002. Engaging Simplifications: Community-Based Resource Management, Market Processes and State Agendas in Upland Southeast Asia. *World Development* 30(2): 265–83.

Lund, C. 1993. Waiting for the Rural Code: Perspectives on a Land Tenure Reform in Niger. IIED Paper, no. 4. *Drylands Network Working Paper.*

McConnell, W. 2002 Misconstrued land use in Vohibazaha: participatory planning in the periphery of Madagascar's Mantadia National Park. *Land Use Policy* 19(3), 217–30.

McNeely, J. and Miller, K. (eds). 1984. *National Parks, Conservation and Development: the Role of Protected Areas in Sustaining Society.* Washington, DC: Smithsonian Institution Press.

MEFT, USAID & CI. 2009. Évolution de la couverture de forêts naturelles à Madagascar 1990–2000–2005. Antananarivo.

Middleton, K. 1999. Who Killed "Malagasy Cactus"? Science, Environment and Colonialism in Southern Madagascar (1924–1930). *Journal of Southern African Studies* 25 (2): 215–48.

Middleton, K. 2009. From Ratsiraka to Ravalomanana: Changing Narratives of Prickly Pears in Dryland Madagascar. *Études Océan Indien. Plantes et Sociétés,* 42–43, 47–83. Paris: INALCO.

Middleton, K. 2012. Renarrating a Biological Invasion: Historical Memory, Local Communities, and Ecologists. *Environment and History* 18: 61–95.

Middleton, K. n.d. Red Prickly Pear and the World Wide Fund for Nature: Rural Poverty and Invasive Species in Dryland Madagascar. Unpublished report.

Middleton, K. 2013. Of Trees and the Political Imagination in "Neo-Colonial" Madagascar. In *Wild Things: Nature and the Social Imagination*, eds. W. Beinart, K. Middleton and S. Pooley. Cambridge: White Horse Press.

Mittermeier, R. 1988. Primate Diversity and the Tropical Forest: Case Studies from Brazil and Madagascar and the Importance of Megadiversity Countries. In *Biodiversity*, eds. E. Wilson and F. Peters, 145–54. Washington, DC: National Academy Press.

Montagne, P. and A. Bertrand 2006. Histoire des politiques forestières au Niger, au Mali et à Madagascar. In *L'État et la gestion locale durable des forêts en Afrique francophone et à Madagascar*, eds. A. Bertrand, P. Montagne and A. Karsenty, 54–83. Paris: L'Harmattan.

Mortimore, M. 1998. *Roots in the African Dust: Sustaining the Sub-Saharan Drylands.* Cambridge: Cambridge University Press.

Murphy, C. 2000. Global governance: poorly done and poorly understood. *International Affairs*, 76(4): 789–804.
Myers, N. 1986. Tropical Deforestation and a Mega-Extinction Spasm. In *Conservation Biology: the Science of Scarcity and Diversity*, ed. M. Soulé, 394–409. Sunderland, Massachusetts: Sinauer Associates.
Nicoll, M. and O. Langrand. 1989. *Madagascar: Revue de la conservation et des aires protégées*. Gland, Suisse: World Wide Fund for Nature.
Olson, D. and E. Dinerstein. 2002. The Global 200: priority ecoregions for global conservation. *Annals of Missouri Botanical Gardens*, 89, 199–224.
Paolina, A., Jarvis, A. and Reus-Smit, C. 1998. *Between Sovereignty and Global Governance: The United Nations, State and Civil Society*, London: Macmillan.
Peluso, N. 1992. *Rich Forests, Poor People*. Berkeley, CA: University of California Press.
Phillips, L. 1986. Madagascar: the Central Highlands. In *Land tenure issues in river basin development in sub-Saharan Africa*, ed. P. Bloch, 116–25. University of Wisconsin-Madison Land Tenure Center Research Paper, No 90.
Poupon, J. 1957. L'aspect particulier du pâturage dans une région semi-aride de Madagascar. Dactylographié, Tuléar 3 janvier 1957.
Pronk, C. and S. Evers. 2007. The complexities of land access in Southeast Madagascar. *Taloha, revue scientifique internationale des civilisations* 43(18).
Raintree, J. ed. 1987. *Land, Trees, and Tenure: Implications for Agroforestry Research and Development*. Madison: University of Wisconsin Press.
Rakotomalala, D. 2008. La création des aires protégées dans le sud et sud ouest de Madagascar: approche et méthodologie. *Malagasy Nature* 1, 168–81.
Randriamampianina, J.A., Solosieva, and J. Rajaonarison. 2002. Rapport de Mission dans le Sud (synthèse) (du 26 novembre au 6 décembre 2001 et du 21 août au 4 septembre 2002). FOFIFA DRA Antananarivo/ FOFIFA centre régional du Sud et Sud-Ouest Toliara/WWF Fort Dauphin/ANGAP Direction Régionale de Toliara. Unpublished Report September 2002.
Randrianarijaona, P. 1983. The erosion of Madagascar. *Ambio* 12, 308–11.
Rauh, W. 1995 and 1998. *Succulent and xerophytic plants of Madagascar*. 2 vols. Mill Valley: Strawberry Press.
Sandlund, O., P. Schei and Å. Viken (eds). 1999. *Invasive Species and Biodiversity Management*. Dordrecht, Netherlands: Kluwer Academic Press.
Sandron, F. (ed.). 2008. *Population rurale et enjeux fonciers à Madagascar*. Paris: Karthala.
Schroeder R. and K. Suryanata. 1996. Gender and Class Power in Agroforestry Systems: Case Studies from Indonesia and West Africa. In *Liberation Ecologies: Environment, Development, Social Movements*, eds. R. Peet and M. Watts, 188–204. London: Routledge.
Scoones, I. (ed.). 1995. *Living with uncertainty: new directions in pastoral development in Africa*. London: Intermediate Technology Publications.
Scott, J. 1998. *Seeing Like a State: how certain schemes to improve the human condition have failed.* New Haven: Yale University Press.
Shepherd, G. 1991. Communal Management of Forests in the Semi-Arid and Sub-Humid Regions of Africa. London: Social Forestry Network ODI.
Soarimalala, V. & Raherilalao, M. 2008. Pression et menaces dans la région forestière sèche malgache. *Malagasy Nature* 1, 157–61.
Stanfield, D. 1995. *Insecurity of Land Tenure in Nicaragua*. University of Wisconsin-Madison Land Tenure Center Research Paper No 120.
Sundar, N. 2000. Unpacking the "Joint" in Joint Forest Management. *Development and Change* 31(1): 255–79.
Sussman R., G. Green, L. Sussman. 1994. Satellite Imagery, Human-Ecology, Anthropology, and Deforestation in Madagascar. *Human Ecology* 22, 333–54.
Sussman R., G. Green, I. Porton, O. Andrianasolondraibe, J. Ratsirarson. 2003. A Survey of the habitat of *Lemur catta* in Southwestern and Southern Madagascar. *Primate Conservation* 19, 32–57.

Système d'Alerte Précoce. 2002. Zone SAP: Données Structurelles concernant la Sécurité alimentaire. AEDES/EU.

Tengö, M., K. Johansson, F. Rakotondrasoa, J. Lundberg, J.-A. Andriamaherilala, J.-A. Rakotoarisoa and T. Elmqvist. 2007. Taboos and Forest Governance: informal protection of Hot Spot Dry Forest in Southern Madagascar. *Ambio* 36(8), 683–691.

Teyssier, A. 1998. Front pionnier et sécurisation foncier à Madagascar: le nécessaire recours à l'Etat. In *Quelles politiques forestières pour l'Afrique rural? Réconcilier pratiques, légitimité et légalité*, ed. P. Lavigne Delville, 583–601. Paris: Karthala.

Thomas, P. 1996. Place, Person and Ancestry among the Temanambondro of Southeast Madagascar. Unpublished Doctoral Thesis, Department of Anthropology, London School of Economics and Political Science, University of London.

Wells, M. & Brandon, K. 1992. Planning for Peoples and Parks: Design Dilemmas. *World Development* 20: 557–70.

World Bank. 1990. Report No 8348-MAG Staff Appraisal Report Democratic Republic of Madagascar Environmental Program March 19, 1990, Agriculture Sector Operations Division South Central and Indian Ocean Department Africa region.

World Bank. 2003. Report No 26106. Madagascar - Rural and environmental sector review, 2 vols.

Wright, P. 1997. The Future of Biodiversity in Madagascar: A View from Ranomafana National Park. In *Natural Change and Human Impact in Madagascar*, ed. S. Goodman and B. Patterson, 381–405. Washington and London: Smithsonian Institution Press.

Young, O., ed. 1997. *Global Governance: Drawing Insights from the Environmental Experience.* Cambridge, MA: MIT Press.

# PARENTING THROUGH BOOM AND BUST IN A NORTHERN MALAGASY MINING TOWN

Andrew Walsh[1]

In early 1996, Ambondromifehy was a small village little different than a dozen others bordering the national highway running through Madagascar's northernmost province. Comprised of several clusters of houses set back from the road, it was home to about 400 people making their living from growing rice and raising cattle. By early 1997, however, everything had changed. Following the discovery of sapphires in its immediate vicinity, Ambondromifehy had exploded into a bustling, chaotic mining and trading boomtown of around 15,000 people.

In the first years of the boom, Ambondromifehy displayed many features common to artisanal mining communities around the world (Biersack 1999, Cleary 1990, De Boeck 1999). It was a place conspicuously dominated by young men and their interests, where newcomers had to both rely on and be wary of reciprocal relations with fellow fortune hunters and where longtime residents' sense of propriety had been drowned out by the urgent inclinations of their newest neighbors (Walsh 2012a). To complicate matters further, most of the sapphires that people were flocking here to dig were to be found within the boundaries of Ankarana National Park, a local conservation area from which all local residents were largely restricted, and to which miners could only gain access illegally. Things rarely stay the same for long in communities built on artisanal mining, however. A diminishing local supply of sapphires and new discoveries of gemstones elsewhere in Madagascar led many foreign buyers and Malagasy prospectors away in the early years of the new century, and by 2003, the population of the town had fallen to less than a third of what it had been at the height of the boom. More recently, in June 2008,

[1] The research on which this chapter is based was supported by a standard research grant from the Social Sciences and Humanities Research Council of Canada. My greatest thanks are due to the people of Ambondromifehy without whose cooperation this research could not have been completed. I am also grateful to the editors of this volume for their comments on an earlier draft.

local officials estimated Ambondromifehy's population at 4000, less than a third of what it was 10 years earlier but more or less the same as what it was five years later when I first started hearing assurances of the town's impending collapse. Over the same period, conservation and police efforts at keeping miners out of Ankarana National Park dropped off, partly due to the fact that the urgency of the perceived crisis of the initial boom had passed, but also to the futility of the task of trying to keep prospectors from the potential fortunes that might be dug from the ground in the region.

One feature of life in Ambondromifehy that has remained constant over the years is something that I expect will be quite familiar to anyone who has spent any amount of time in a small-scale mining boomtown: this is a place in which people must learn to live with constant uncertainty. No one I spoke with in 1999 could ever have known that that the initial boom would end up lasting as long (or as short) as it did, that foreign buyers would come and go in waves, that new discoveries elsewhere would come when they did and draw people away, or that the decline of the local trade would in fact make life better for many among the tenacious third who have stayed behind over the long term (Walsh 2012b). Running concurrent with all that is uncertain about life in this place, however, have been trends that anyone might have predicted. Young men and women worked and met. Children were born. Parenthood brought new responsibilities. People aged. People died. A place in which few prospectors intended to stay more than a couple of weeks or months at the start has become a place in which many of them have now lived for more that a decade, and this combination of time's passing and people's persistence has made Ambondromifehy a very different place today than it was when it first boomed.

In this chapter I present Ambondromifehy's trajectory over a decade as the backdrop for a discussion of a range of topics that strike me as relevant to the themes of this volume. I hope that by taking a longitudinal approach I might highlight how all relationships, whether intergenerational or those connecting people with the land they occupy, have been differently shaped and influential at different times, as people have adapted to both the uncertainties and the inevitabilities of life here. As I have been doing in my research, I tack between two intertwined groups in Ambondromifehy as I proceed. I begin with the longtime residents who lived here before the boom, move on to the much larger group that has come since, and end with a discussion of how, 10 years into the boom, the latter group has come to resemble the former. Since the idea of parental responsibility is central to all that follows, I begin with a few words about the role of parents in other Malagasy communities I have known.

### *A Place with Parents, 1999*

In Malagasy villages of the sort I lived and worked in prior to doing research in Ambondromifehy, it is common to refer to older residents as the 'parents' (*ray-amandreny*) of a community. These are people who consider themselves, and are considered by others, to be responsible for just about everything that goes on around them, and as such, they are people who tend to take an interest in, and often attempt to manage, the behavior of their juniors and neighbors. Parents are the ones to whom one is meant to report incidences of, and/or seek advice on, significant transgressions of social conventions or taboos (Walsh 2002a), and, as a group, they are the ones generally called upon to deliberate over and settle the disputes that can arise in such situations (Gezon 2006). They are also generally the people asked to preside over the initiation of new building projects, the practice of life cycle rites, and requests for blessing through invocations of ancestors and local land spirits. Broadly speaking, a community's parents are those responsible for assuring the conditions necessary for individual and communal reproduction and prosperity in a particular place. Unsurprisingly, then, the parents of a community generally play key roles in mediating people's relationships with the land from which community members make their livings. I will have more to say on this shortly.

A community's parents are also the ones to whom a newcomer is meant to present themselves upon arriving in a place, and so it is that I initially met a couple of Ambondromifehy's original parents on my first visit to this town in 1999. Zama was around 75 at the time, living in one of the town's few well-built, plank-floored houses. His younger brother Tombo was around 60. Over the next few months these men shared their version of the story of what Ambondromifehy had once been and the place it had become.

Originally established in the early 1900s by young cattle herders from Madagascar's East coast, Ambondromifehy grew slowly through the colonial era as these men settled here with their wives and had children, clearing and planting rice fields, building houses and, as time went on, establishing sacred sites, including tombs, at which invocations of God, ancestors and local land spirits might take place. By the time Zama and Tombo had children of their own, the place had become about as large as it would ever get in pre-boom years. In fact, to hear others tell it, by the time of the boom, Ambondromifehy had been in decline for some years. In keeping with a common pattern in the region, many of the village's young men and women had been drawn away by more attractive work opportunities in nearby towns, leaving elders like Zama and Tombo with a lessened burden of parental responsibility that, although perhaps easier to bear, was not something they ever would have wanted. When young people leave a community like the one Ambondromifehy used to be, they take much of the vitality on which such places rely for their reproduction with them. Without youth, who will provide the labor necessary to maintain fields, irrigation ditches and public facilities? Who will take in harvests and keep track of cattle? Who will do the heavy lifting when the hardwood coffins of entombed ancestors need replacing? And most important, perhaps, without youth, how can men like Zama and Tombo realize the power that comes with parental responsibility? No wonder, then, that as young people (and young men especially) began arriving in Ambondromifehy in droves in the early months of the sapphire boom, Zama and Tombo were among the first to welcome them, a decision that Zama quickly came to regret.

Before proceeding any further with Zama and Tombo's attitudes towards their community's new arrivals, it bears mentioning that these elders were not the only ones concerned with the demographic changes that the sapphire boom had brought to the region in which they lived. Also concerned were conservation workers charged with managing Ankarana National Park, the protected area near Ambondromifehy in which the region's sapphires were most readily available to prospecting miners. Although first established as a "special reserve" under French rule in 1956, the park's international reputation as a site of distinctive ecosystems and extraordinary biodiversity did not develop until the 1980s, following several high-profile research projects involving foreign and Malagasy scientists and conservationists (cf. Wilson, Stewart and Fowler 1988). In the early 1990s, the park also came to figure significantly in the plans of conservationists intent on promoting development through ecotourism in the region; located just off a national highway connecting two already popular tourist hubs, and home to both the spectacular landscapes and charismatic species for which Madagascar is known around the world, it promised to become a significant draw for wealthy foreign travelers and, thus, to provide opportunities for sustainable development to local inhabitants (Freudenberger 2010). The discovery of sapphires within the park's boundaries, then, posed a double threat to conservationists – not only was sapphire mining within park boundaries deemed to be destructive, unsustainable and thus a threat to the park's ecosystems and biodiversity, in that mining left some parts of the local landscape denuded of trees and riddled with pits, it threatened the park's reputation as a pristine and protected area among foreign ecotourists. It wasn't just mining inside the park that posed a problem, however. As significant over the long term have been the expected effects of a significant influx of new residents settling in the immediate vicinity of the park – new residents who, like the longtime residents they joined, have come to pose a threat to local ecosystems by virtue of nothing more than the pressures exerted by living in the region.

From Zama's perspective, the problems brought on by the sapphire boom were *not* simply the result of an influx of thousands of new arrivals. "Many people is a good thing", he often repeated, citing a proverb that stresses the advantages of a growing population. His complaints, rather, were directed at the specific, disrespectful and fundamentally irresponsible, behaviors of these newcomers. Of particular concern to him were the problematic ways in which these 'visitors' (*vahiny*) were misusing local land. Miners were clogging up local water sources by sieving through sacks of clayey dirt in search of sapphires. They were destroying carefully planned, hand-dug irrigation systems with their random 'dropping' of pits. Most egregious of all, perhaps, they were transgressing local land taboos (including the regional taboo that proscribes digging the earth on Tuesdays) in their relentless pursuit of gems. Also of concern to Zama were the ways in which his new neighbors were misusing local sacred sites at which invocations of local ancestors, local land spirits and God had previously occurred. In past years, such sites were places at which the dynamic networks of mutual responsibility on which this community relied for its reproduction were played out for all to see. The large tree outside of Zama's house, for example, was once a place at which residents of Ambondromifehy could publicly seek blessing in support of their own or their children's future projects. Under the guidance of Zama, this invocation site's primary caretaker, and accompanied by other community parents, residents would approach the tree, invoke local land spirits, ancestors and God, inform them of a specific goal for which blessing was being sought and, finally, promise to return to the site with a specified sacrifice once this goal had been achieved. With their eventual return and the fulfillment of promises made, instigators of this process would not only highlight the efficacy of the forces from which they sought blessing and reinforce the sanctity of the site at which they did so, they would also assert themselves as responsible members of the community in their own right. Put another way, in addition to being places at which blessing could be sought, sites like this one were places in which categorical 'parents' made and proved themselves.

In the early months of the boom, a number of settlers who had taken to calling Zama 'father' or 'grandfather' after being granted permission to settle near his house asked him to help them secure profitable positions in the then still nascent local sapphire trade by allowing them to seek the blessing of local ancestors and land spirits at the tree outside his house. Initially, Zama was as accommodating with them as he would have been with anyone he called 'child' or 'grandchild'. Over the next couple of months, however, as more newcomers arrived, the direction of things to come became evident. Newcomers stopped seeking Zama's permission to settle, and even those who called him father or grandfather began neglecting him. Several of the young men who had sought blessing and made promises at the base of his tree were even brazen enough to ignore Zama's entreaties that they should live up to the promises they had made, a failure to act that was not only disrespectful to Zama but indicated a profound disregard for the community of which he was a parent. When promises made at sites like this are not lived up to, it is not taboos but, rather, the sites themselves that are said to be broken. Even more problematic than the forgetfulness of those Zama had initially welcomed were the very public and deliberate taboo transgressions of a congregation of evangelical Christians occupying the banks of a marshy sacred lake at which ceremonies involving cleansing and bathing had taken place in previous years. Here, in direct contravention of Zama's warnings, this congregation had built a church and begun to purposely defile the lake, intent on proving that Jesus was a far more powerful force on the earth than the 'satanic' land and ancestral spirits that they admonished longtime residents for worrying about (Walsh 2002b).

The sorry situation faced by Zama two years into the boom was particularly evident late in 1999 when he tried to spread news of a dream he had had in which a land spirit associated with the abovementioned lake appeared to him in the form of a crocodile and informed him of its intention to leave the region as a result of having been disrespected and defiled (cf. Lambek 2002). In previous years, reports of such a dream would likely have rippled through the whole community, causing much debate, finger pointing and consternation. In 1999, however, amidst the chaos of the boom, it seemed as though I was one of very few people to take any interest. Not even the people who called Zama 'father' or 'grandfather' were concerned about what this dream might mean or what role they might play in rectifying the situation, failing to see, as Zama did, that their success in this place was, like his, dependent on their behaving responsibly, not just with regard to community parents but to the land itself. A key element of Zama's largely unheard commentary on his dream was that the much discussed dwindling supply of easily accessed sapphires in the region was not the result of over-mining, as some assumed, but was, rather, a result of the disrespectful and irresponsible acts of visitors to the region. His message was a familiar one in Madagascar: sustainable prosperity comes to those who listen to their parents, or more specifically, to those who live according to the "ethos of growth" (Keller 2008) advocated by parents – an ethos that demands responsible interchange not only with kin and neighbours, but with local landscapes as well. As noted earlier, however, few were listening.

Although Zama was the most senior of Ambondromifehy's original parents, and his complaints were shared by many others, it would be a mistake to over generalize based on his experiences alone. Consider Tombo, for example, Zama's younger brother. Years before the boom, Tombo had established himself apart from the rest of the family in a newer house about 500 meters down the highway from Zama. Although he shared his brother's concerns about the behavior of newcomers, and was troubled by their taboo transgressions in particular, Tombo was also young and well enough to be able to engage in some of the opportunities offered by the local mining boom. By 1999 he was already sponsoring the efforts of some of the miners who had settled near him, providing them with rice from his nearby fields and other kinds of support in exchange for a share of the profits of their mining. Sometimes he even worked alongside them in their pits. Tombo also gained renown among his new neighbors as a diviner and spirit medium, offering them advice on everything from what direction to follow in extending their mining pits to how they might deal with the misfortunes brought on by their neglect of their own, usually distant, elders and ancestors. Perhaps because of these closer associations with the town's newest residents, Tombo was not as quick as Zama to give up on the young men and women living around him. With the characteristic crude flair that made him so popular among many of his much younger neighbors, he once explained to me that "young people are like assholes ... even though they might be dirty and causing you problems, there is no getting rid of them". This improvised proverb was not a product of apathy. Rather, it captured a sentiment that was doubtless shared by Zama and all of Ambondromifehy's other original parents: the only thing worse than a community overrun by irresponsible young people is a community with no young people at all. A community without children has no hope of a future, let alone a prosperous or sustainable one.

Having offered some sense of how two of Ambondromifehy's original fathers sought to manage the new opportunities and uncertainties brought by sapphire mining boom of the late 1990s, I turn next to the prospectors themselves and the opportunities and uncertainties that life in this community presented to them.

## *A Place without Parents*

One of the ways in which people in Ambondromifehy explained what was so unique about this place back in the late 1990s was by describing it as one in which the vast majority of inhabitants had "met when grown" – i.e., it is a place in which people have made lives for themselves and with one another despite the fact that they have no roots, and generally no parents, here. "Meeting when grown", I was frequently reminded, is in many ways a good thing. It means that people are exposed to different ways of thinking about and doing things, different *fomba* or customs, and it affords them the opportunity to forge new relationships with people they might never have met otherwise. For many of the miners and traders I met in 1999, however, a community in which people "meet when grown" was also attractive for what it lacked: parents, or, more precisely, parents in whose eyes one feels obliged to behave responsibly.

Elsewhere I have discussed how, in the early years of the boom at least, the largely young, and dominantly male population of Ambondromifehy might be included among those in the world who Sophie Day et al. describe as "living for the moment", i.e., people whose apparent "freedom and autonomy stand in opposition to transcendental values associated with a variety of institutions that organize long-term social reproduction and, simultaneously, produce hierarchical relationships" (Day et al. 1999, 2 cited in Walsh 2003, 291). Engaging in conspicuous consumption without thought to the sort of responsible investments that deliberative elders would expect them to make, and transgressing taboos with little concern for consequences, were only two of many sorts of 'daring' enabled, or at least made much easier, by life without parents in this place. Also significant was the fact that access to a livelihood in this place, or more precisely access to the land that would make earning a livelihood in this place possible, was not mediated by parents of any sort. As noted earlier, in the boom's early years, mining commonly occurred within the boundaries of Ankarana National Park, where miners worked illegally, and defiantly (Baker-Médard 2012), under threat of being arrested, detained and/or fined by the national police who had been called in to help conservation workers protect the place. Others mined on rocky, unfarmed land that local residents had no good reason to gain title to previously. As such, they were under no obligation to share what they took from local land with men like Zama and Tombo. Not that all of the miners and traders I met in 1999 ignored their responsibilities, however. In fact, many migrant prospectors sent what savings they could amass home to distant parents and places as remittances, sometimes even requesting that water carrying ancestral blessing be sent back to them in return. Such responsible behaviour mattered little to elders like Zama and Tombo, however. Indeed, it wasn't even perceptible locally. However well Ambondromifehy's prospectors did, and however responsibly they may have invested their earnings, Ambondromifehy and its original parents remained largely neglected.

From Zama's perspective the fact that Ambondromifehy had become, in his own words, a "place without parents" was problematic for a number of reasons. As he saw it, a place without parents is a place in which the irresponsible behavior of youth, however much expected, can go unchecked to the detriment of all. To whom was he meant to address his concerns about the behavior of his new juniors if not to their parents? And without these other parents to back him up, who would help him rein in these youth and force them to channel their energy into the accomplishment of communal, rather than individual, goals? (cf. Bloch 1999).

In 1999, I attended a meeting that highlighted the gap between the priorities of Ambondromifehy's original parents and those of its newest inhabitants. The meeting had been called to discuss the outcome of an earlier visit by representatives of the Malagasy Ministry of Education during which Zama, Tombo and a number of other community parents had been assured of government support for the much needed expansion of the town's single public primary school. Since a condition for this assistance was that Ambondromifehy's parents would have to contribute money and labor to the expansion project themselves, this meeting had been called in order to gather interested parties to confer on the matter and undertake some unfinished work on the existing school building as a show of good faith. Based on my past experiences in communities in the region, meetings involving collective work projects like this are not unusual, and although people grumble about them, most will attend anyhow well aware of the stigma associated with not doing so. In Ambondromifehy in 1999, however, only 10 men showed up to the meeting in question, all of them over the age of 40 and all but two of them people who had lived here before the boom. Unable to offer anything but his presence, Zama sat in the shade, watching as nine fellow parents (only a few of whom had school age children) half-heartedly undertook about an hour's work on the school's long neglected single building.

To give a sense of how much things changed in Ambondromifehy in the years after the boom, let me flash forward to another meeting that occurred at this same place in June 2008. The reason for this more recent meeting was the same as for the earlier one – the town's one public primary school was too small to meet the demands of the town's population, and while government support had been offered for an expansion, it would only be forthcoming if the community's parents contributed money and labor themselves. This time, however, the meeting took place in one of the school's three cement floored classrooms (built several years earlier with community labor and support), and was presided over not by the community's original parents but by the school's three teachers and about 20 members (men and women) of its "Parents' Association", an organization comprised of the parents of children attending the school, all of whom were post-boom newcomers to this place. In addition to discussing the expansion project, there was much talk from the teachers about how some parents were not paying their monthly school fees, and much concern from parents about the finances of the school itself. This was followed up by a lengthy discussion over the merits of different fundraising schemes that ended with plans to undertake what would be the school's second annual community fundraising dance. Finally, there was much tut-tutting from all sides about the parents who failed to show up at the meeting, leading one teacher to suggest that those who had come should spread the word that the children of parents who failed to live up to their commitments to the school would be sent home. None of this was ideal, of course, but neither was much of it particularly new to me. Over the years, I have attended similar meetings at which many of the same issues and tensions surrounding schooling and parental responsibilities came to light. Never before in Ambondromifehy, however.

## *The Return of Parents, 2008*

What happened in the years between the meetings just described to effect Ambondromifehy's apparent transformation from being a place without parents to a place with, at very least, a "Parents' Association"? First and most obviously it is worth mentioning that many of the people who had come here originally in 1999 had stayed and had children here. But there is more to it than that. As mentioned in the introduction, the slowdown of the local sapphire trade had brought significant changes to Ambondromifehy, and for many of the people who stayed behind, these changes made this not just a better place to be, but a better place to be parents. Take, for example, the fact that by 2008 the population was less than a third of what it had been in 1999. To those I interviewed in 2006 and 2008, this drop in population was more likely to be discussed in positive than negative terms. Houses that were once packed together, crammed up as close as possible to the side of the road, had become more widely spread, making everyone less susceptible to the fires that had previously regularly ravaged large sections of the town. Some who stayed had even taken over the abandoned lots of former neighbors to either expand their own homes or courtyards, or to plant fenced gardens of fruit trees, manioc and greens. This population drop had also, by all reports, made Ambondromifehy a much safer place to be. Drawing a contrast with the early years of the boom, one miner told me in 2008 that you can "now sleep with a bag of money as a pillow and not have to worry about it." Indeed, many of the miners and traders I have spoken with in post-boom years indicated that they much preferred living in Ambondromifehy than in other, still booming, gemstone mining towns in southern and central Madagascar. For one thing, the facilities available to parents here were better than in these other places; in addition to there being a greatly expanded public primary school in town in 2008, there were seven private primary schools operating at capacity.

The decline of the local trade was also accompanied by a decline in police and conservation agency oversight of the sapphire-rich conservation area that was once the exclusive domain of stealthy, risk-taking young men, meaning that by 2008 one encountered almost as many women and children as men working and living within the boundaries of Ankarana National Park. Where disputes over claims to particular pits inside the park were once settled by brawls, by 2008 they were more likely to be deliberated over by a group of senior men who passed judgments of the sort that they might come to in regulating land disputes among farmers. Conservation workers in the region continued to object to mining within the park's boundaries, but their objections fell on deaf ears. Although miners were not making nearly as much from their finds in 2008 as they had been in the early years of the boom, the opportunities represented by the local sapphire trade were clearly attractive enough to keep many of them involved.

I should be clear, however. Ambondromifehy has not become, nor will it ever be, just a larger version of the place it was before the boom. As of 2008, sapphires remained the main business in town, and there is no question that the distinctive sorts of work associated with this commodity have irrevocably changed the local economy and local social relations (see Walsh 2004; Walsh 2009). What is more, none of even the most settled of Ambondromifehy's post-boom arrivals I spoke with in 2008 had cut themselves off entirely from home regions or communities. In fact, the most popular sorts of voluntary association in town at this time were not parents' associations, but those that provided people hailing from the same home region with support in times of crisis, or with money to help cover the cost of sending a corpse home for a proper burial when a death occurs. As one of the teachers I interviewed in 2008 complained, people here are often more inclined to provide for ancestors in distant places than for the children living in their own houses, quick to contribute money if a body needs transporting to a distant tomb, but nowhere to be found when their child's school is in need of renovation.

However much they might be seen, in retrospect, to have lived, and thrived, with decline in recent years, the day to day existence of Ambondromifehy's remaining residents is, in fact, much more about living with constant uncertainty (cf. Ferguson 1999). The circumstances they faced in 2008 illustrates this well. In May of that year, Madagascar's then President Ravalomanana declared a moratorium on the export of rough gemstones, supposedly in an effort at both cracking down on smuggling and promoting Madagascar's emerging gemstone cutting and polishing industry. The immediate effects of this policy in Ambondromifehy were what you might expect. First, foreign gemstone buyers stopped buying anything but the best of what was on offer in town; before long, they stopped coming to town altogether. This left Malagasy traders with the option of either sitting on their stock in anticipation of the eventual return of foreign buyers, or taking advantage of the lull and lack of competition to buy up good quality stones at low prices from desperate miners. The same questions were on everyone's minds. Why had the President done this? When would the trade open up again? When would the foreign buyers be back? If and when they did come back, what sorts of prices would they be offering? When the basis of local livelihoods is as precarious as it is in Ambondromifehy, it makes perfect sense that people living here hedge their bets by retaining connections with distant places,

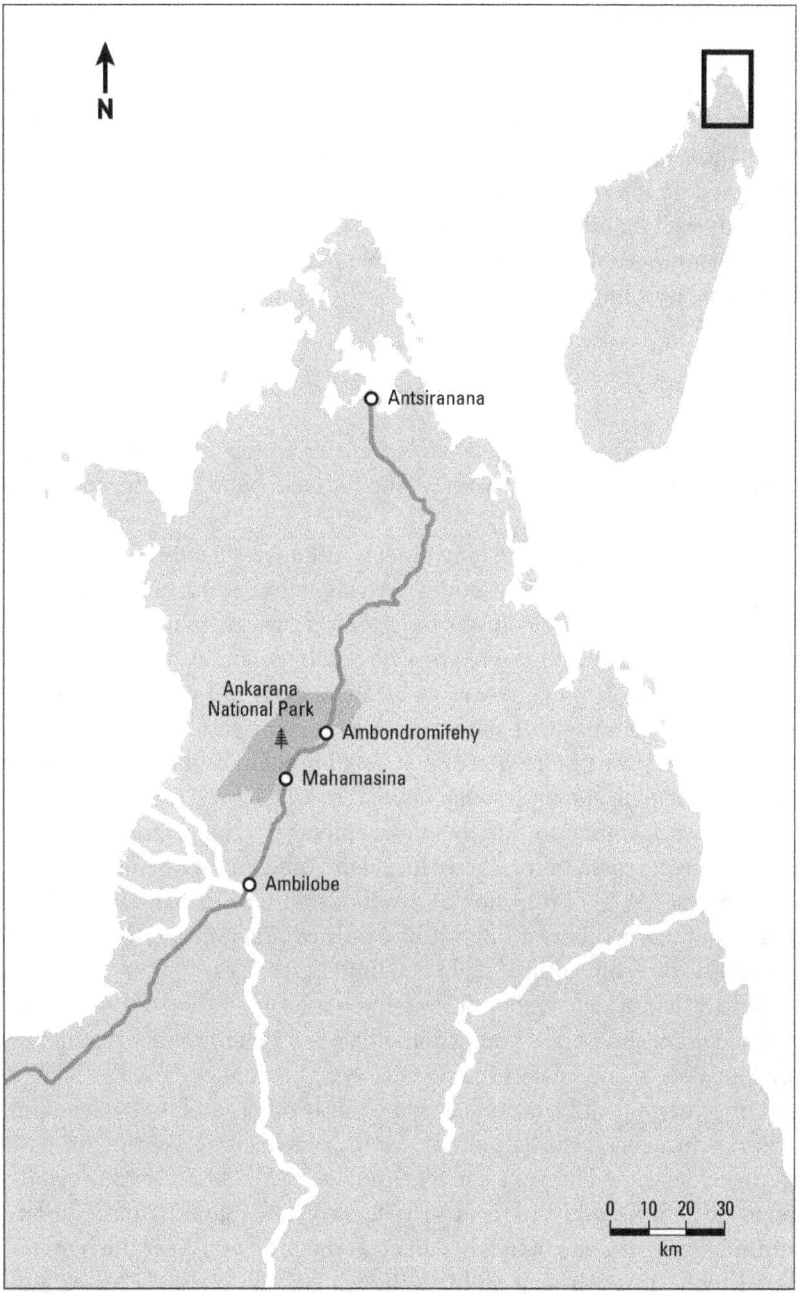


Map 1. Northern Madagascar
Originally published in: *'Made in Madagascar: Sapphires, Ecotourism and the Global Bazaar'*, © University of Toronto Press Higher Education Division 2012.

however burdensome keeping those connections up might seem at times. Under these circumstances it also makes sense that some of the town's new parents are just as reluctant as the government and NGOs to invest too much in developing local public facilities that may be unneeded in a year's time. Although most of the miners and traders I interviewed in 2008 expressed no interest in leaving Ambondromifehy, they also understood perfectly well that it was not *their* interests, but those of distant consumers, gemstone wholesalers and government officials, that will ultimately determine how long they will stay.

## *Conclusion*

To end, I should return briefly to Ambondromifehy's original parents. What became of Zama and Tombo, the two men who introduced me to the history of this place?

Zama died in 2004 after a long illness. Although surrounded by young people, many of them post-boom arrivals who continued to call him 'grandfather' until the end, it was reportedly a struggle to raise the money necessary to get him to the nearest hospital, 100 km away. When I last spoke with him in 2003, he was as bitter as ever, still regretting having accommodated those first prospectors who came to his door. Tombo's death, in 2005, was more of a shock, leading many I have spoken with since to speculate about just what it was that killed him. Had Tombo been too cozy with Ambondromifehy's newcomers? Had he been too cavalier about allowing them to transgress local land taboos that he was meant to be enforcing? When I returned to Ambondromifehy in 2006, I found his house and many others around it had burned down in the town's most recent fire. By 2008, nothing had been built in its place.

Shortly before I left Ambondromifehy in 2008, I searched out Zama's 23 year old grandson Bora, a young man I had first met nine years earlier on my first visit to the town. I found him living permanently with his wife and one-year-old child, as well as two of his brothers and their wives and children, in an oasis of shade amidst family rice-fields two kilometers east of town. I reminded him of a conversation we had had four years previous during which he told me (as a 19 year old) how much better Ambondromifehy was in 2004 than it had been in the pre-boom years he remembered living through as a child. Although he had certainly heard and understood the complaints of his own parents and grandparents over the years, he was also unapologetically drawn to the life, "*la vie*", that the sapphire trade had brought to this place. In 2004, Ambondromifehy was a place with bars, video parlors, regular organized fighting matches, well-stocked shops, and a daily market at which one could buy foodstuffs like vegetables, dried shrimp and fresh meat, products that in previous years one would have had to travel 30 kilometers to find. In 2008, I asked him if he felt the same. Of course things are better now than they had been before the boom, he assured me, listing many of the same attractive features of life here. If anything, he was even more appreciative now than before. Living permanently outside of town means that he doesn't have to put up with any of the inconveniences of Ambondromifehy. Water at his new home is relatively plentiful, houses here can be spread as far apart as you like, there is room for kids to play, gardens are easily kept and need not be fenced in, cattle can be grazed and watched over nearby, etc. With the town so close, however, he could still enjoy all of its advantages: a constant market for the rice coming out of his fields, shops at which to stock up on essentials, the everyday availability of all kinds of food, and even the occasional video or fighting match to watch.

For several years, a couple of men who had contracted fictive kin relationships with Tombo had been trying their own luck at growing rice with their families, returning a third of their harvest to Bora and his family in return for their use of the land. And these newcomers to responsible relationships with local land and landowners were not alone. The more the local sapphire trade declines, it seems, the more visitors who have decided to settle here seek opportunities to farm. And the more they do so, the more young men like Bora realize their own status as parents of the community developing around them, and as owners and protectors of the arable land on which this community is becoming increasingly dependent. In 2008, Ambondromifehy was nothing like the community that Bora's grandparents once knew. In at least one respect, however, it had become the kind of place that I suspect Zama and Tombo would have hoped for – a place in which their grandchildren could envision prosperous futures for themselves, their own descendents, and anyone else who would come to see them as parents through time.

## *Bibliography*

Baker-Médard, Merrill. 2012. Conflicting Treasures: contrasting use governance in two artisanal gemstone mining sites in Madagascar. *Journal of Political Ecology*, 19: 221–237.

Biersack, Aletta. 1999. The Mount Kare Python and His Gold: Totemism and Ecology in the Papua New Guinea Highlands. *American Anthropologist*, 101(1): 68–87.

Bloch, Maurice. 1999. 'Eating' Young Men among the Zafimaniry. In *Ancestors, Power and History in Madagascar*, ed. Karen Middleton, 175–217. Leiden: Brill.

Cleary, David. 1990. *Anatomy of the Amazon Gold Rush*. Iowa City: University of Iowa Press.

Day, Sophie, Evthymios Papataxiarchis, and Michael Stewart, ed. 1999. *Lilies of the Field: Marginal People Who Live for the Moment*. Boulder: Westview Press.

DeBoeck, Filip. 1999. Domesticating Diamonds and Dollars: Identity, Expenditure and Sharing in Southwestern Zaire (1984–1997). In *Globalization and Identity: Dialectics of Flow and Closure*, ed. Birgit Meyer and Peter Geschiere, 177–209. Oxford: Blackwell Publishers.

DeBoeck, Filip. 2001. Garimpeiro Worlds: Digging, Dying & 'Hunting' for Diamonds in Angola. *Review of African Political Economy*, 28(90): 549–562.

Ferguson, James. 1999. *Expectations of Modernity: Myths and Meanings of Urban Life on the Zambian Copperbelt*. Berkeley: University of California Press.

Freudenberger, Karen. 2010. *Paradise Lost? Lessons frm 25 Years of USAID Environment Programs in Madagascar*. Washington, DC: International Resources Group.

Gezon, Lisa. 2006. *Global Visions, Local Landscapes: A Political Ecology of Conservation, Conflict, and Control in Northern Madagascar*. Lanham, MD: Altamira Press.

Keller, Eva. 2008. The banana plant and the moon: Conservation and the Malagasy ethos of life in Masoala, Madagascar. *American Ethnologist* 35(4):650–664.

Lambek, Michael. 2002. Nuriarty, the Saint, and the Sultan: Virtuous Subject and Subjective Virtuoso of the Post-Modern Colony. In *Post-Colonial Subjectivities*, ed. Richard Werbner, 25–43. London: Zed Books.

Walsh, Andrew. 2002a. Taboos, Responsibility and the 'Freedom to Do Otherwise" in Northern Madagascar. *Journal of the Royal Anthropological Institute* 8(3): 451–468.

Walsh, Andrew. 2002b. Saving Souls, Preserving Bodies: Religious Incongruity in a Northern Malagasy Mining Town. *Journal of Religion in Africa*, 32(3): 366–392.

Walsh, Andrew. 2003. 'Hot Money' and Daring Consumption in a Northern Malagasy Sapphire Mining Town. *American Ethnologist* 30(2): 290–305.

Walsh, Andrew. 2004. In the Wake of Things: Speculating in and about Sapphires in Northern Madagascar. *American Anthropologist* 106(2): 225–237.

Walsh, Andrew. 2009. The Grift: Getting Burned in the Northern Malagasy Sapphire Trade. In *Economics and Morality: Anthropological Approaches*, ed. Kate Brown and Lynne Milgram, 59–76. Lanham, MD: Altamira Press.

Walsh, Andrew. 2012a. *Made in Madagascar: Sapphires, Ecotourism and the Global Bazaar*. Toronto: University of Toronto Press.

Walsh, Andrew. 2012b. After the Rush: Living With Uncertainty in a Malagasy Mining Town. *Africa* 82(2): 235–251.

Wilson, Jane M., Paul D. Stewart, and Simon V. Fowler. 1988. Ankarana: A Rediscovered Nature Reserve in Northern Madagascar. *Oryx* 22(3): 163–171.

# DISCOURSE, DEVELOPMENT AND LEGITIMACY: NATURE/CULTURE DUALISM OF MINING ENGAGEMENTS IN BIODIVERSITY OFFSETTING AND CONSERVATION IN MADAGASCAR

Caroline Seagle[1]

In January 2013, chief executive Nyfanja Rakotomalala of Rio Tinto/QMM, along with nearly 200 other mine workers, was found trapped inside the mine shaft by hundreds of Malagasy protesters demanding better compensation for land and mining impacts and more job opportunities for local Antanosy people rather than outsiders.[2] This was not the first time local groups in Madagascar's Southeast Anosy region organized protests against the mine; in 2009, protesters set up road-blocks preventing Rio Tinto/QMM trucks from entering the Mandena processing plant (L'Express Madagascar 2009). In these protests, which occurred fieldwork, grievances ranged from inadequate compensation for land, broken promises and an overall lack of transparency – to negative impacts on fishing communities who had lost access to fisheries due to a dam built by the company, and the displacement of people from their ancestral land because of Rio Tinto/QMM's acquisition of a mountain for a rock quarry. For many people displaced, the replacement land provided by the company was not seen as having equal value to the land lost; as one woman stated, "*Roambifolo*! Twelve ancestors are the real, legal owners of this land. QMM did not give us equal compensation for this." The resettlement

[1] Many thanks to Editorial Board of this volume, Gwyn Campbell, Sandra Evers and Michael Lambek, and the contributing authors. Deepest gratitude to Dina Navalona Rasolofoniaina (MSc., University of Antananarivo), who helped me tremendously in the field and was kind enough to translate Malagasy interviews into French. Our working partnership was facilitated through part of a wider research collaboration which pairs Masters' students in the Netherlands and Madagascar, organized by Sandra Evers (supervisor of MSc research) in partnerships with the VU University Amsterdam, ICCO and the *l'Institut de Civilisations/ Musée d'Art et Archéologie* at the University of Antananarivo). Many thanks also to the *l'Institut de Civilisations* in Antananarivo, Madame Rondro, the mayor of the commune of Ampasy Nahampohana, and the various informants who were kind enough to share their views and stories with me between January-March 2009.

[2] The Telegraph. 2013. "Madagascar forces free 200 Rio Tinto hostages." http://www.telegraph.co.uk/finance/newsbysector/industry/mining/9798009/Madagascar-forces-free-200-Rio-Tinto-hostages.html, Accessed 18 January 2013.

area was said to be host to land entirely unsuitable for farming, and the woman claimed that villagers had not seen compensation payments since 2006.

These contestations over land – while of utmost importance to Malagasy people, who have a strong relationship to the land for both subsistence purposes but also for maintaining social, ancestral and kinship relations – remain tertiary to Rio Tinto's two other objectives: profit earnings from ilmenite extraction (estimated at billions during the life of the mine), but also a less obvious goal for a multinational mining company: to set a "benchmark" in "green mining" and become a leader in sustainable development (Prinsloo 2009; see also McNeilly 2000). Only two years prior to the 2013 protests, chief executive Rakotomalala was photographed planting seedlings of fast-growing species, Acacia and Eucalyptus, as part of the Rio Tinto's pledge to rehabilitate the formerly biodiverse, 6,000 hectare (ha) littoral forest sitting atop the mineral deposits and targeted for strip mining.[3] The company's tree nursery near the Mandena deposit contains seedlings collected from the rare littoral forest which will be destroyed for mining, and forms part of the ecotourism circuit set up by the company; the circuit traverses a conservation zone set aside within the mining perimeter and from which local people are heavily restricted and charged fines for entering. In a Rio Tinto publication, following the company's receipt of the prestigious Nedbank "green mining" award,[4] Rakotomalala stated, "We are delighted that QMM and its local and global partners have been recognised for protecting and conserving the unique biodiversity of the mining sites while building capacity for better use of the natural resources by the local community" (Rio Tinto, date n/a, "Biodiversity Strategy and Net Positive Impact"). These same words were used by QMM's Manon Vincelette, former head of the company's biodiversity programme who, following news of the award, highlighted the company's actions to preserve biodiversity and build capacity for "better use" of natural resources.[5] While the species used for rehabilitation were not endemic to the littoral forest destroyed by the company, Rio Tinto noted that the plantations would serve as "compensation measures

[3] Rio Tinto. 2001–2009. "Official launching of the rehabilitation of the Mandena mining site." http://www.riotintomadagascar.com/english/bioNewsrehab.asp, Accessed 18 January 2013.

[4] See Rio Tinto. 2009. "Rio Tinto wins Green Mining Award." http://www.riotinto.com/media/18435_group_news_18700.asp, Accessed 18 January 2013.

[5] Ibid.

essential to achieving the net positive impact of mining on the environment" as well as compensation to local people.[6]

Steps taken by the corporate sector, particularly some mining companies, to "preserve" the environments in which they operate and bring sustainable development are part of a broader trend in the market economy. The embedding of "nature" and ecosystem services in global market chains has changed the way conservation is practiced, imagined and financed by international actors, sparking a phenomenon recently coined "green grabbing" – the large-scale acquisition of territory for conservation (Fairhead, Leach and Scoones 2012). However, the emerging green economy[7] has also created opportunities for the corporate sector – notably multinational mining companies – to reverse decades of land access barriers arising from negative perceptions of mining's impact on people and the environment. Corporate adherence to voluntary guidelines such as CSR (Corporate Social Responsibility), consciously designed in terms of global sustainable development aims and ideologies (Kapelus 2002), as well as the restructuring of mining laws to welcome foreign investment (see Sarrasin 2006) and an explosion of "mining-conservation" partnerships (a phenomenon that has not been met without debate) has meant that, over the past ten years, large-scale mining companies have come to play a crucial role in shaping sustainable development and nature conservation.[8] In other parts of the World, mining and conservation have been entangled with each other, and in Papua New Guinea West (2006) referred to mining and conservation as "strange bedfellows" – but rarely have these two actor groups worked so closely together as in Madagascar, a biodiversity hotspot (Myers et al. 2000) and mineral-rich nation that has witnessed a surge in mining projects over the past twenty years.

Mining and conservation appear to unite around recent trends in green economy (e.g. payments for environmental services, REDD,[9] conservation finance, biodiversity offsetting). As such, many scholars have pointed out how neoliberalism – as development ideology emphasizing free markets, individuation, privatization, and economic growth (Harvey

---

[6] Rio Tinto. 2001–2009. "Official launching of the rehabilitation of the Mandena mining site." http://www.riotintomadagascar.com/english/bioNewsrehab.asp, Accessed 18 January 2013.

[7] See "Greening the World Economy," IUCN. http://www.iucn.org/what/tpas/greeneconomy/, Accessed 25.09.12.

[8] See Mining, Minerals and Sustainable Development (MMSD) programme, http://www.iisd.org/mmsd/, Accessed 18 January 2013.

[9] Reducing Emissions from Deforestation and Degradation.

2005) – increasingly influences the way both conservation and sustainability are delimited, defined and valued by international actors (Büscher et al. 2012; Igoe and Brockington 2007; Sullivan 2010; Fortwangler 2007; McDonald 2010; Castree 2008). This chapter takes this discussion to Madagascar, where two multinational mining projects led by Sherritt/Ambatovy[10] and Rio Tinto/QMM[11] aim to set standards for "green" mineral extraction through partnerships with conservation NGOs, the creation of protected areas within mining sites and "biodiversity offsetting" schemes – compensation for biodiversity loss caused by mining impacts through the designation of land for biodiversity conservation in other parts of Madagascar (Ten Kate, Bishop and Banyon 2004, see also Mail & Guardian 2009). It draws upon three months of ethnographic fieldwork[12] carried out in 2009 specifically on the Rio Tinto/QMM ilmenite mine (Southeast Madagascar), which unearthed various impacts on local populations with a high reliance on land and natural resources (Seagle 2012). In Madagascar, particularly in rural areas, land is more than an economic asset but also a source of livelihood, food security, social-power relations, conceptions of growth and kinship, and broadly speaking, identity and ancestral relations (Evers 2006; Keller 2008). Impacts documented in 2009 included land dispossession and involuntary resettlement, the decimation of wetland reeds used for weaving, vast livelihood shifts, the removal of tombs, lost access to natural resources, lost access to biodiversity, and job termination following brief three year contracts with the mining company. Many of these and other impacts on rural populations are well-documented (Harbinson 2007; ALT and PANOS 2009, Seagle 2009). Kraemer (2012) for example has shown how new forms

[10] Ambatovy is a joint venture between Sherritt (40%), Sumitomo and Korea Resources (each hold a 27.5% stake), and the contractor for the project, SNC-Lavalin (5% interest). In the BBOP, the project is referred to as "The Ambatovy Project." For the purposes of this chapter, "Sherritt/Ambatovy" will be used to refer to the project and project operator. In partnership with Conservation International, Forest Trends and Wildlife Conservation Society, it is implementing a pilot project for the BBOP (Business and Biodiversity Offset Programme).

[11] Rio Tinto, one of the World's largest multinational mining companies, is operating a 6,000 hectare ilmenite (titanium dioxide, a whitening pigment) mine in Madagascar's Southeast (Anosy region). Mineralized zones encompass a rare littoral forest that will be removed through a process of strip mining with the use of a wet dredge. Rio Tinto is working with its subsidiary, QMM (*QIT – Quebec Fer et Titane – Madagascar Minerals*) to mine and implement the various sustainable development and conservation projects initiated by the company. For this purposes, "Rio Tinto/QMM" is used to refer to the mining company. "QMM" is often used locally to refer to the company.

[12] Fieldwork carried out was part of the author's Masters research at the VU University Amsterdam.

of social inclusion and exclusion have been created as a result of the project's influence in the region.

With reports of growing inflation, land remains in high demand due to continued foreign investments in the region and economic opportunities for expatriates who buy or lease land as concessions for ecotourism or tourist resorts; combined with Rio Tinto's own conservation actions (which include ecotourism and protected areas co-managed by local groups), these dynamics reflect what Büscher and Davidov (2013) have termed the "ecotourism-extraction" nexus. This chapter critically analyzes the "mining-conservation nexus," specifically focusing on some of the legitimization strategies used within the discourses and media used by Sherritt/Ambatovy and Rio Tinto/QMM surrounding nature conservation and biodiversity offsets. Drawing upon recent theories exploring resource ontologies, embodiment of environmental perception and land as heritage, it suggests that mining-conservation approaches to compensation and nature conservation may uphold the nature/culture dualism prominent in Western environmental thought that does not reflect processual realities and valuations of land use in many rural parts of Madagascar. By conceptually separating – rather than reconciling – humans from the environment, mining and conservation run the risk of undervaluing (and thereby under-compensating) Malagasy people affected. Land may be seen to have both material and intangible value as local heritage (Evers and Seagle 2012) and daily activities structured around land-use shape not only economic opportunities but also identity, traditional environmental knowledge (TEK) and social-cultural relations.

The sections that follow will unpack these theoretical underpinnings and provide a Madagascar overview. It then goes on to use two examples, analyzing media surrounding mining's engagement in conservation and "biodiversity offsetting," to describe how Rio Tinto/QMM and Sherritt/Ambatovy approach nature conservation and local communities.

## *Madagascar: Brief Historical and Contemporary Context*

Madagascar is well-known for its rare biodiversity; the fourth largest in the world has often been referred to as a giant Galapagos Islands. The majority of Madagascar's population live in rural areas and have a high dependence on land for livelihood purposes (Sarrasin 2006). In Southeast Madagascar (as in many other parts of the country), these activities range from rice and other forms of crop cultivation (sweet potato, manioc), fishing, the collection of non-timber forest products – such as fibres and medicinal plants, tree felling and charcoal production. During the colonial period (1986–1960), the French controlled vast swathes of land and forests and regularly curbed local landscape burning practices for subsistence, called *tavy* (or "slash and burn" cultivation) (Jarosz 1993). Control over forests (particularly through colonial logging and conservation campaigns) was also a means of pushing Malagasy people into forced wage labour regimes and enforcing taxation (Sodikoff 2005; see also Campbell 2005). Current assessments of deforestation, which tend to pinpoint Malagasy land users as the main culprits of environmental degradation, thus often fail to capture the complex history of these contestations over land and forests (Kull 2000; Jarosz 1993).

Reeling from the lingering effects of international debt, failed Structural Adjustment Policies (SAPs) initiated the 1970s and pressure from the World Bank to modernize through economic liberalization whilst protecting the environment, Madagascar witnessed the growth of transnational networks of environmental governance and foreign interests in biodiversity conservation in the 1970s and 80s, and today, international conservation NGOs play a powerful role in governance (Horning 2006; Duffy 2006; Corson 2010). This trend was intensified by debt relief mechanisms such as "Debt for Nature (DfN)" swaps (see Deacon and Murphy 1997)[13] which created pathways for increased conservation finance. Local dependencies on land and forests have often come into conflict with the practices and goals of conservation NGOs seeking to enclose protected areas, and disputes between people and parks have intensified over the past 20–30 years (see Keller 2008; Ferraro 2002; West, Igoe and Brockington 2006; Simsik 2002; Harper 2002, among others). In the midst of growing neoliberal solutions to biodiversity loss in Madagascar and elsewhere (Brockington et al. 2008), many conservation projects follow a "payments" approach wherein local populations are offered compensation (and alternative livelihoods training) to make up for lost access to natural resources. This has also meant, in many cases, the *transfert de gestion* (management transfer) of the protected area to people living on park peripheries (see Hockley and Andriamarovololona 2007). In such a case the conservation zone is usually patrolled and "managed" by villagers, though under strict

[13] See WWF (World Wildlife Fund). 2008. "Monumental Debt-for-Nature Swap Provides $20 Million to Protect Biodiversity in Madagascar," (June 11, 2008). Available from: http://worldwildlife.org/press_releases/monumental-debt-for-nature-swap-provides-20-million-to-protect-biodiversity-in-madagascar-wwf-announces, Accessed 22 January 2013.

rules, and the sense of "ownership" is intended to create incentives to conserve natural habitats. Walsh (2005) and others (see Harper 2002; Liverman 2004) have shown that the perceived benefits arising from conservation or ecotourism projects, while well-intentioned, often fail to trickle down to people living near park perimeters and sometimes create burdens or friction within communities, who are sometimes resorted to policing each other.

### *Mining, Neoliberalism and Biodiversity Conservation*

Recent years have witnessed heightened demand for land in Madagascar, both for conservation but also for mineral extraction. Part of a larger wave of "land grabbing" (see Burnod et al. 2011) and "green grabbing" (see Corson 2012; Neimark 2012; Seagle 2012) in Madagascar, the rush for land has literally led to overlapping boundaries of protected areas and mineral concessions. Cardiff and Andriamanalina (2007) found that mining concessions overlapped with a large percentage of forests set aside as existing or planned conservation zones in Madagascar, a phenomenon which increased between 2005 and 2006.

As a means of addressing business risks and growing land access barriers, the extractive industry began to position itself at the forefront of a putative "global transition to sustainability"[14] in 1999, a shift coinciding with the "mainstreaming of sustainability into the marketplace" (Adams and Jeanrenaud 2008: 32) and the explosion of various think-tanks on corporate ethics and sustainability. As "nature" became an arena for increased profit-making, sustainability and CSR discourses were streamlined into a corporate legitimization strategy (see Littlewood and Wells 2000 on the "Global Mining Initiative; see also Tsing 2000 and Kapelus 2002). Sharing the common goal of land access, conservation NGOs increasingly "partner" with multinational mining companies to negotiate land access, environmental performance and mitigation strategies. These partnerships have also meant the down-toning of criticism (in broader media) against mining actors by partner NGOs, and diverted attention to notions of opportunity, consensus and partnership. Corporate engagement in 'sustainability' discourses, institutionalized in Corporate Social Responsibility (CSR) and various other corporate think tanks (e.g. the MMSD, TEEB,

14 Rio Tinto. 2013. "Our Strategy." http://www.riotinto.com/ourapproach/our_strategy.asp, Accessed 20 January 2013.

BBOP, ICMM),[15] as well as remediation techniques such as biodiversity offsetting and mining-conservation partnerships may signal a new way in which mining and conservation actors "grasp hold" of power – and more generally, an emerging political economy of extraction in Madagascar and elsewhere. While some studies have analyzed how mining and conservation affect local land users in the same locality, and how actors involved might make similar promises of "development" in contexts of both conservation and mineral extraction (West 2006), rarely have these two actor groups worked so closely together.

As a form of capitalist expansion, the coupling of markets and conservation may conceal the need for neoliberalism to fulfil its project of reaffirming unequal power relations – indeed "uneven development" – and maintain the hegemony of elite classes (Harvey 2006; Igoe, Neves and Brockington 2010). But there is more to the story; regulatory mechanisms such as CSR and the EITI (Extractive Industries Transparency Initiative) have shaped multinational discursive and mediation techniques (Kapelus 2002). Catchphrases such as "sustainability," "participation," "accountability," and "local community" increasingly make up the chorus of what Barry (2004) calls "ethical capitalism," and sustainable development has become an industry of its own (Welker 2009). Scholars are nevertheless beginning to question the "win-win" scenarios inherent to these "good governance" mechanisms and discourses (Hall 2011; Borras et al. 2011; Li 2008). While on the one hand this type of re-regulation institutionalizes companies' claims to protect society and nature, on the other it brings up important issues of power and corporate paternalism (Rajak 2008). For example, many EIAs (Environmental Impact Assessments) and other studies carried out under the banner of accountability (and verified by scientific studies) are produced – and therefore *defined* – by companies themselves. In her analysis of an EIA administered by a gold mining company in Peru, Li (2009: 219, italics in original) argues that "the *form* of the documents produced for EIA (i.e. their required components, as established in legal frameworks) and the *process* of making them public (participatory meetings and public forums) can take precedence over their *content*." The company's emphasis on collaborative rhetoric such as "participation" often over-shadowed the grievances of local people (Li 2009). While focused on the "improvement" of life and ecosystems (in the vein of

[15] Mining, Minerals and Sustainable Development (MMSD; Business and Biodiversity Offset Programme (BBOP); The Economics of Environment and Biodiversity (TEEB); The International Council on Mining and Metals (ICMM).

Foucault's concept of biopower, see Genel 2006), CSR may mark a fundamental shift in the way companies grasp hold of power and access land. Many international environmental NGOs (IENGOs) as well as the World Bank now see the corporate sector as playing a crucial role in biodiversity protection. Various new partnerships have emerged to date; for example, Rio Tinto has entered into partnerships with Conservation International (CI), WWF Australia, IUCN (International Union for the Conservation of Nature), BirdLife International, Kew Botanical Gardens, and the World Conservation Society. Sherritt/Ambatovy is working closely with IUCN, Wildlife Conservation Society (WCS), Conservation International (CI), Missouri Botanical Gardens, and UNICEF.[16]

The following section will explore some of the theoretical foundations for this chapter before delving into a document analysis of mining's approach to conservation, offsetting and sustainability.

### *Theoretical Frame: Resource Ontologies and the Embodiment of Environmental Perception*

A common critique of political ecology is its prioritization of ideological or social constructions of nature over actual material realities; while some have called for a more "political" political ecology (Büscher 2009: 48–49), others have emphasized the need for more "ecology" in the discipline (Walker 2005). Scholars are beginning to look at the materiality of resources and mutuality of human-environment interactions in an attempt to dissolve the long-standing "society-nature" dualism that has long shaped Western environmental thought. In Western epistemology, nature and culture are generally thought to be separate and distinct entities; conversely, in other parts of the world, the boundaries are much more complex and blurred. An attempt to conceptually dissolve this distinction has involved a critical examination of the "corporeal" (bodily) or "body geographies" as a means of understanding the "biophysical and social character of material existence" (Bakker and Bridge 2006: 17). This work transcends constructivist approaches to nature by suggesting a dialectical relationship between human bodies and land (Demeritt 2002). Central to this discussion is the concept of embodiment (Scheper-Hughes and Lock 1987) which, echoing Latour's (1993) call to collapse the nature/culture

16 See Ambatovy website, "Partnerships," http://www.ambatovy.com/docs/?p=505, Accessed 20 January 2013.

divide, rejects Cartesian dualisms separating mind and body (see also Merleau-Ponty 2005[1945]). As Bakker and Bridge (2006: 20–21) point out: "Work on embodiment offers new approaches to analysing the shaping of social relations by material conditions, without falling prey to environmental or biological determinism." Similarly, studies on materiality often emphasize the "hybrid" – an analytical tool which allows resource anthropology to bypass "subject/object" and "nature/society" binaries by highlighting the relational, distributed and dialectical basis of human-environment interactions (Swyngedouw 2004; Whatmore 2002). In other words, the qualities of "things" are not inherent to their materiality or to social perceptions of them, but derived from associations and interstices found between bodies and nature, which change over time (see also Castree 1995).

Ingold (drawing upon Lefebvre) has pointed to the importance of relational ontologies and the embodiment of environmental perception in his work on "meshwork" and "lines" (2007: 80). In attempting to understand Aboriginal conceptions of landscape, he notes that the "lines" – or traces people inscribe in their immediate environs – constitute not a network but a "meshwork" or interwoven web of social-environmental interrelations that traverse past, present and future (Ingold 2007: 80). Relating landscapes to "zones of entanglement," Ingold notes that the environment is "that which surrounds" (2007: 103) – similar to how the Malagasy define the word, "environment": *tontolo iainana* (literally "the world in which we live"[17]). This ideology gives some insight into how, theoretically, one might approach Malagasy perceptions of the environment and "ethos" of life itself (see Keller 2008) – wherein dichotomies between nature/culture and static representations of culture do not hold up. This work also provides insight into how resources are spatialized – how historical and geographical spaces (nature) are produced through practice and process (Lefebvre 1991[1974]; see Smith 1984). The way in which resources (e.g. water, forests, land, biodiversity, minerals) "come into being" through these processes has been termed 'resource ontologies' (Richardson and Weszkalnys, forthcoming). Resource ontology refers to how a resource's materiality is distributive or 'translated' (see Latour 1993) and how social-cultural relations are organized around particular aspects of a resource. Webb Keane (2006) has referred to this process as "bundling," wherein a set of qualities specific to certain resources or objects "bring them into

[17] This quote was related to me during fieldwork and data collection in 2009.

being." An underlying hypothesis in this chapter is that local 'resource ontologies' provide an important proxy for understanding contestations over, and valuations concerning, land, labour, wealth, and livelihoods in Madagascar.

In Madagascar, land use signals the "metaphysics of the mundane... people seek in land not just material satisfaction but also power, wealth and meaning (Shipton 1994: 348)." These theories help unpack why [notably cash] compensatory mechanisms proposed by Sherritt/Ambatovy and Rio Tinto/QMM – through their development and conservation strategies – may or may not acknowledge the importance of land in rural Malagasy settings, or what Roymans et al. (2009) have termed "landscape ontologies."

### *Embodying Land-use in Madagascar: Conceptions and Valuations of Land*

As Madagascar's rural population is primarily subsistence-based, relying heavily upon land for both material purposes (food security, income, medicinal plants) as well as for existential or ontological reasons (connecting to ancestors, upholding ancestral knowledge, labour values, inheritance), many foreign-led projects are viewed locally as *mivarotra tanindrazana*, or "selling off the land of the ancestors" – a Malagasy proverb with roots in the French colonial period. Tombs are quintessentially the property of the ancestors, a type of inheritance passed down to future generations. Land (*tanindrazana*, "land of the ancestors") anchors many Malagasy people in customs (*fomba*) and taboos (*fady*) laid out by the ancestors; in turn, notions of "ownership," growth and kinship, as well as access to power and social relations, are connected to land. As one informant, Angeline, explained: "Land is inheritance; it is the *donneur de vie* (giver of life); the question is: what will they *do* afterwards, when their land is gone? Cultivating the land is their heritage: it is a symbol of family value. The ancestral ties between parents and children are not simply economic."

The above quote points not only to the conceptual importance of land as a living organism (e.g. "donneur de vie") and providing for the continuation of life, but also to the value of land-labour relations, of everyday interactions with the environment. By inquiring as to what people will *do* when they are unable to access land, the quote conveys the central importance of natural resources in structuring everyday lives. In one case, a fishing village was requested by Rio Tinto/QMM to stop fishing for three years, purportedly as a conservation measure but also due to the nearby construction of an international port to ship minerals.[18] Villagers mentioned that they had refused the monetary compensation offered in order to maintain access to fishing grounds; while the access to these fishing grounds was ultimately cut off (see ALT and PANOS 2009) the resistance to compensation points also to the value placed on livelihood security through human-environment interactions. One proverb related to me by another informant, JP (Feb, 2009), explains this phenomenon well: "*Hamboly akondro, hamboly tsonjo* (to plant bananas is to plant *tsonjo*)" *Tsonjo* is a type of tuber, similar to manioc, and a staple food source (thus important for everyday subsistence). The shoots growing at the base of the plant symbolize that the banana plant is constantly in a state of reproducing itself; the banana plant is thus a metaphor for the continuation of life, and never death. JP and I discussed this proverb in relation to Malagasy conceptions of kinship, inheritance and lineages; the saying is often used at marriage ceremonies or when one wishes for someone else to have children. Importantly, Keller (2008) has described the metaphor of the banana plant in relation to the Malagasy "ethos" of life itself; she suggests that the concept is linked to the Malagasy notion of "moving forward" and becoming "rooted" in new land, if necessary (Keller 2008: 651–652). Mostly, however, it underlines the importance of land in structuring social relations (termed the *fihavanana*) between both the living and the dead. As another Malagasy proverb explains: "*Ny tany vadiben'i Zanahary mihary ny velona manotrona ny maty* ("The earth is the first wife of God: it cares for the living and embraces [wraps] the dead") (Fox 1990: 30). Thus instead of being something essential or innate, "nature" is constituted by the very interactions between people ("we") and life itself. For these reasons, the nature/culture dualism that has long shaped Western environmental thought is inherently problematic in the Madagascar context (Simsik 2002).

These examples underline the importance of viewing land as local heritage (Evers and Seagle, 2012; Cormier-Salem and Bassett 2007); as

---

[18] Further eviction has occurred near Port Ehoala, where 440 hectares of land called "Ehoala Park" was acquired by Rio Tinto/QMM following the Malagasy government's designation of the land as "public utility" (property of the State). The eviction process was purportedly administered by the State, but it is unknown how many people were actually displaced. Rio Tinto/QMM attempted to sell the land to interested investors in 2012 at the International Fair of Madagascar (FIM, see http://www.riotintomadagascar.com/english/newsfim.asp, Accessed 23 January 2013). Rio Tinto/QMM claim that the eviction process was carried out in adherence to World Bank (international) regulations and that adequate compensation was offered, although the company does not detail what – or how much – the compensation was or amounted to. See http://www.riotintomadagascar.com/english/newspark.asp, Accessed 23 January 2013.

Roymans et al. (2009: 339) state, "heritage is always the dynamic work of people, with processes of cultural transmission and the construction of values and identities being inextricably bound up with one another." Lost land access or sudden environmental changes may thus threaten various aspects of people's livelihoods and value systems.

### I. *Strategies of Legitimization: Biodiversity Offsets and the Role of Nature and Culture*

Before first analyzing how the nature/culture dialectic is represented in biodiversity offset discourse, it is first necessary to recall the tenets upon which Western environmentalism and concepts of "sustainability" are based (see also Luke 2005). Having emerged from post-Enlightenment thinking (emphasizing individuality, scientific reason and rational thinking), French Cartesianism (dualisms of mind/body and society/nature) and Romanticism (nature as having inherent value, or "pristine wilderness" to be protected for aesthetic appreciation, in the vein of John Muir), Western environmentalism aimed to balance Romanticist tropes such as the "wilderness ideal" with economic approaches to natural resource management, such as "maximum sustainable yield."[19] These two imperatives gradually became the basis upon which global conservation agendas were based (Van Dyke, 2008). The export of Western environmental thought to global contexts re-emphasized the separation of humans from nature and, needless to say, created complications in parts of the world where human interactions with nature do not adhere to solely aesthetic or econometric standards, such as in many parts of Madagascar.

Today, global conservation efforts are generally shaped around notions of consensus and cooperation with regard to sustainability, biodiversity conservation and climate change (Van Dyke 2008; Büscher 2009). Fletcher has noted that conservation has traversed phases spanning "fortress conservation" (see Brockington 2002) to "flexible" conservation (ICDPs[20] and CBNRM[21]) involving local participation, and now to "fictitious"

[19] 'Sustainable yield' refers to a process wherein maximum extraction is balanced against equal rates of renewal in, for example, logging. It is the primary ideology beyond "sustainable use" discourse omnipresent in conservationist rhetoric.

[20] ICDP stands for Integrated Conservation and Development Projects (see Gezon 1997).

[21] CBNRM stands for Community-Based Natural Resource Management (see Dressler et al. 2010).

conservation – which we see in the context of value circulation in global markets for environmental services (e.g. species banking and markets for carbon and biodiversity, see Büscher 2012[22] and Sullivan 2012). Fictitious conservation is largely responsible for what a growing body of scholars are referring to as "green grabbing" (Fairhead, Leach and Scoones 2012) which may also underpin the nexus of mining and conservation.

Fictitious conservation may be particularly important for analyzing partnerships forged around biodiversity offsetting, defined as "conservation actions designed to compensate for the unavoidable residual impacts on biodiversity caused by mining and processing (Ten Kate, Bishop and Banyon 2004: 1)." As mentioned above, biodiversity offsetting allows for companies to remediate negative impacts on biodiversity by conserving elsewhere. Central to biodiversity offsetting discourse is the promise of a "Net Positive Impact (NPI)" on biological diversity and "No Net Loss (NNL)" of species (Ibid). While biodiversity loss caused by development can occur in one location, it can be "offset" through actions to preserve biodiversity in another location. Rio Tinto (2007:1, italics added) note that an offset "takes place away from the impact site and normally takes the form of averted disturbance or ecological restoration." *Averted disturbance* refers to "reducing existing external impacts such as land clearing by other forest users, lessees or owners" (Rio Tinto Biodiversity Offset Strategy 2008: 1), thus while mining is permitted, local subsistence use is not; this echoes what Luke (1997)'s discussion of how certain conceptions of "right" and "wrong" natural resource use are embedded in sustainability and conservationist discourses.

Markets for biodiversity abound, and are on the rise. Mining companies can buy and sell biodiversity "credits" (called "biocredits") to offset negative impacts; this normally works through a third party or "BioBank" that handles the transactions – and major BioBanks have emerged in England and Australia to help provide biodiversity offset credits sought by developers.[23] Many conservation NGOs and organizations (notably IUCN) and mining companies worldwide are experimenting

---

[22] *Büscher* coined the term "fictitious conservation," see *Büscher*, Bram. 2012. Nature on the Move: The Emergence and Circulation of Fictitious Conservation and Liquid Nature. Paper presented at the Nature Inc. conference at the Institute of Social Studies The Hague (June/July 2012).

[23] See Levitt, T. 2010. "What is biodiversity offsetting and how would it work?" The Ecologist (9 June 2010). Available from: http://www.theecologist.org/News/news_analysis/499256/what_is_biodiversity_offsetting_and_how_would_it_work.html, Accessed 23 January 2013.

with biodiversity offsets. Biodiversity offsets strongly resemble fictitious conservation, or the circulation of environmental services in market chains, due to the necessity of categorizing, classifying and monetarily valuing ecosystems (or parts therein) based on specific criteria, then relying on markets to provide the playing field. However, we don't know what fictitious conservation actually "looks like" on the ground; what are the actual impacts of biodiversity offsetting on local Malagasy people, who also have claims to land and biodiversity? What modes of inclusion and exclusion are involved? Fieldwork conducted in 2009 suggests that models of nature conservation in Madagascar are gradually moving back towards "fortress conservation" – more [human] exclusion from protected areas.[24] This would bypass previous models which emphasized human inclusion and "participation." While this remains to be seen, a report on biodiversity offsets emerging from the IUCN-Rio Tinto partnership warns:

> [f]aced with development of a mine and development of biodiversity offsets, there is a real risk that local communities may face a 'double whammy' of negative impacts from both initiatives (e.g. if a community is dependent upon forest resources, and its access to forest is reduced through mining-caused deforestation and the implementation of a 'fortress-style' protected area) (Temple et al. 2012).

This suggests that there might be a feedback mechanism wherein "fictitious conservation" links back to "fortress conservation" but bypasses "flexible conservation" (human inclusion). Implementation processes, access restrictions and contestations over valuation methods in offset design are less clear.

In short, likely due to emerging (and profitable) markets, "biodiversity loss" is increasingly seen as both a business risk but also "part of the case for access to land (business opportunity)" by mining companies (Rio Tinto/QMM, 2004, "The Biodiversity Partnership Program": 1). In turn, the TEEB (Economics of Ecosystems and Biodiversity) report claims that "biodiversity loss and ecosystem degradation" presents "both serious risks to business, as well as significant opportunities" (Elmqvist et al. 2010: 8). Biodiversity offsetting brings into question not only the scale of "green grabbing" (Fairhead, Leach and Scoones 2012) for minerals, but also how

---

24 Interview conducted with a representative from WWF in Fort Dauphin; cross-referenced with a PhD field researcher (Antonie Kraemer) who had heard the same point from interviews she conducted.

compensation is negotiated and valued both globally and locally. Mining-conservation partnerships seem to be heavily mobilized around offsets; the BBOP states that "a company may work in partnership with an NGO to implement a biodiversity offset, and the NGO may attract additional funding to expand its work in the area (BBOP 2009a: 14)." Similarly, Barry notes that companies attempting to act responsibly often try to make social links with actors holding ethical "authority," such as NGOs and academics (Barry 2004: 203).

### *Biodiversity Offsets, (de-)valuation(?) and Local Communities*

While biodiversity offsetting is primarily global compensation, it is premised on "payments for environmental services": paying locals for lost access to forests and land. "Unsustainable" and "illegal" land use activities are regularly cited in the BBOP (Business, Biodiversity and Offset Programme) handbook as barriers to achieving a so-called "Net Positive Impact" on biodiversity (see chart summary on BBOP 2009a: 40). Communities are generally recognized as having high stakes to lose or gain in cases of both mining and conservation, though by being included as "managers" of the offset, the BBOP handbook expects that local communities will have more incentives to participate (ibid: 42). Nevertheless, "poor performance by the developer on other issues may affect how the community works on the offset (ibid: 44)." In turn, while local populations may or may not feel the sense of "ownership" of the offset (as indicated in the BBOP), as Peluso and Ribot (2003: 153) aptly state, having rights to an asset does not always ensure the ability to *access* it.

With regard to local communities and offsets, Ten Kate, Bishop and Banyon (2004: 6) state the following:

> Local communities can use biodiversity offsets to ensure functioning and productive ecosystems during and after development projects, not only with properly rehabilitated project sites, but with additional conservation outcomes outside the project's borders, to support livelihoods and amenity (Ten Kate, Bishop and Banyon 2004: 6).

From what we have learned already about how "rehabilitation" is approached by the Rio Tinto (planting fast-growing tree plantations), the promise of rehabilitated forests – without regard to local uses of biodiversity (which are widespread) or time-scales, remains vague. Also, how local populations would benefit in detail from "additional conservation outcomes outside of the project's borders" is not clear. Lost access to natural resources is an obvious outcome of biodiversity offsetting (BBOP 2009b). Many livelihood alternatives proposed involve the dissuading of subsistence cultivation for staple crops and promotion of market-based livelihoods, organic gardening (with help from local NGOs) and ecotourism. Some local people will be recruited as guards to patrol the offset (ibid). However, "in-kind" or "out-of-kind" compensation to local communities for biodiversity remains an issue and involves a complex process of socio-cultural valuation laid out in the BBOP's "Cost and Benefit" handbook (2009b), which focuses specifically on local communities. While some forms of "in-kind" compensation are proposed in the handbook, such as replacing lost access to particular medicinal plants with plant nurseries supplying the same plant, no mention is made of scale (e.g. the size and capacity of the nursery needed) or the time period necessary to produce viable, harvestable plants (BBOP 2009b: 41). The handbook concedes that "Many socioeconomic offset options can be considered out-of-kind...If valuation is applied appropriately, there is no economic reason why a cash payment should not be considered to offset a local community's loss of fuelwood or access to NTFPs[25] as a result of the project (ibid)." From what we've seen above: the critical question to ask is how cash compensation can adequately make up for non-economic values of land access and biodiversity, and under whose terms. Finally, the BBOP does not adequately deal with traditional environmental knowledge (TEK) and potential effects on knowledge transmission (see Berkes et al. 2000).

What we can observe from these examples thus far is that while the BBOP – as well as Rio Tinto and Sherritt/Ambatovy – acknowledge the need for compensation (both for *in situ* mining impacts and *ex situ* conservation measures), the way in which compensation is valued and administered appears very complex and has its challenges. The BBOP notes that "cash payments" will be used for "loss of cultural sites" as well as opportunity costs due losing access to both "labour and land" (BBOP 2009b: 47). Following preliminary discussions with Malagasy specialists earlier this year, one purported key grievance of local populations affected by the Ambatovy nick mine is a lack of attention to "culture." In their Environmental Assessment (EA), Sherritt/Ambatovy tend to isolate aspects of local "culture" to tombs and "sacred sites" whose value can be adjusted over time; for example, Sherritt notes that "[a]lthough the sacred waterfall at the mine site cannot be relocated, its sacred aspect can be

25 NTFPs=Non Timber Forest Products.

*modified.* This is accomplished by virtue of the people's loss of contact with it – if rituals cannot be conducted there, it is no longer considered sacred by them (Ambatovy EA 2006: 17, emphasis added)."

Likening their biodiversity strategy to their approach to culture, Rio Tinto has proposed to "offset" impacts on culture with "cultural heritage offsets" (2011b: 57). "Cultural offsets" are nearly identical in approach and rhetoric as "biodiversity offsets" (see above) and manifest in the documentation of oral histories, research and publications on tangible cultural sites, construction of museums, and conservation of culturally important landscape features "outside" of the mining sites, to be used by local populations. As heritage is here interpreted in relation to embodied, dynamic uses of land and social-ancestral relations, the static preservation of sites deemed culturally equivalent to other locations could have serious repercussions to people who value land for non-economic (e.g. ancestral) purposes. This would require a valuation method going beyond cash compensation. Nevertheless, 'cultural offsets' are thus targeted to have a "net positive impact on cultural heritage" (Rio Tinto 2011b: 74). What we see here is the use of an 'offset ideology', where there is a tendency to measure, compartmentalize and quantify increasingly immeasurable things (e.g. biodiversity, culture) and "offset" them with their supposed equivalents in other locations. What is at risk here is a disregard for the way in which people embed themselves in a landscape (see discussion above) and the importance of certain locations versus others. Again, the tendency to separate "cultural" from "natural" appears to be problematic in biodiversity offset design.

We can also see, in these dynamics, Scott (1998)'s concept of 'legibility' – wherein state attempts to rationalize, demarcate and label complex social phenomena encompass a means in which bureaucratic structures extend their control over populations. The state (and transnational entities) does this by absorbing local pluralities into hegemonic discourses and legal structures; Scott uses the example of laws surrounding privatization, census, land tenure and allocation, though we can also think about legibility within the context of human-nature relationships and mediations of sustainability. Through media, mining companies can make legible certain aspects of the social and ecological reality and others. In short, cultural offsets and biodiversity offsets appear to uphold the nature/culture dualism by analytically valuing culture and biodiversity separately.

The offset ideology also is premised upon market rationality. With regard to labour, the BBOP document notes that "[t]he opportunity cost of labour can be estimated by reference to market wage rates or the average production per unit of labour" (ibid: 53); however, this may not adequately reflect social realities in Madagascar where most rural labour is unpaid (subsistence) and involves a high expenditure with low returns in production. Moreover, again it is an economic, quantifiable valuation of labour – with no consideration of non-economic livelihood valuations, such as knowledge transmission of certain forms of labour (fishing, farming etc) or the everyday need for food security. There is an underlying risk of undervaluing land and labour in the BBOP, and thus under-compensating affected people.

Polayni (1944) has described how capitalism's tendency to treat land and labour as commodities to be captured in monetary terms points to an inherent fallacy; this is his premise behind his notion of "fictitious commodities" (land, labour and money). Using embodiment as a basis for understanding environmental perception, it is important to consider how (and what kinds of) compensation may or may not capture rural Malagasy people's relationships with land and labour; in Madagascar, land-based activities (e.g. rice cultivation, fishing) cut across all members of the family and generations: men, women, children, adults, and elders. Cash compensation for lost land access brings into question these social dynamics. As one informant, JP, mentioned in 2009 (in relation to individuals impacted or displaced by the Rio Tinto/QMM mining project), "QMM gave them money, and now they are poor."

While the BBOP attempts to render biodiversity "legible" (Scott 1998) through market valuation methods, and while Sherritt/Ambatovy and Rio Tinto/QMM use striking images to mediate the importance of biodiversity they claim to save, these practices may not adequately account for the '*illegibility*' of local uses, values and perceptions of the environment. In 2009, it was found that the Rio Tinto mine decimated stands of an important species endemic to the littoral forest called *mahampy*, a wetland reed used by women to weave mats, baskets and hats (data collected in Mandena forest). *Mahampy* is a source of much needed additional income during periods of scarcity, and is also the *only* species and material employed to wrap deceased ancestors before they are placed in a tomb. At the time of research, the loss of *mahampy* was seen as a major blow to women living on the peripheries of the mine, and more broadly speaking, an impact on the existential security of people (both women and men) living near Mandena, who could not wrap their ancestors in traditional burial shrouds. Many complained that adequate compensation for this loss had not been received, and women had to travel long distances to fetch *mahampy* from forests to the North – where the second site of extraction (St. Luce) is now under way. In short, while the presence of humans is not absent in the BBOP, there still appears to be a separation of nature/culture the overall design, with a priorization of biodiversity values rather than local relationships with the environment.

The next section will analyze some of Rio Tinto/QMM's mediations of conservation in relation to local land uses.

## II. *Strategies of Legitimization: Biodiversity Book and Narratives of Degradation*

While Rio Tinto/QMM tends to downplay the importance of biodiversity to locals, the company took pains to identify the "intrinsic value" of littoral forest biodiversity for the global good; this manifested in a Rio Tinto/QMM publication (published by the Smithsonian Institute) entitled, Biodiversity, Ecology, and Conservation of Littoral Ecosystems in Southeastern Madagascar, Tolagnaro (Fort Dauphin) (Ganzhorn, Goodman and Vincelette, eds. 2007). The book involved the participation of several internationally-recognized biologists and is a very impressive synthesis of over twenty years of scientific research carried out in the littoral forests which Rio Tinto/QMM paradoxically must clear (the mineral deposit covers 6,000 hectares[26]).

The mediation of the biodiversity book, in addition to partnerships formed with environmental NGOs and inclusion of small conservation zones within each mining site periphery, creates the semblance that a multinational mining company is instead an environmental NGO engaged in research and conservation. Before analyzing the tone and narratives in the book, it is first necessary to briefly outline some of the deforestation myths in Madagascar.

### *Discourses of Degradation and Deforestation*

As deforestation in Madagascar is commonly characterised as a direct cause of poverty and inappropriate local land use practices (Gezon 1997: 463), many conservation NGOs tend to characterize rural Malagasy people as either "villains" or "victims" of environmental degradation (McConnell 2001: 17) trapped in self-perpetuating cycle of poverty and

[26] It should be noted that 620ha in total are being set aside as conservation zones *within* each mining site.

"wrong" natural resource use (see Luke 1995). Such characterizations stem from an historical vilification of *tavy*, shifting cultivation or "slash and burn" agriculture, the main agricultural practice on the island. *Tavy* is an ancestral practice wherein small (usually not larger than 1.2 ha) patches of land are burned of their vegetation, allowing the nutrients from the ash leech into the soil and provide a nourishing environment for rice, manioc and other staple crops to grow. Typically, *tavy* was a nomadic practice (Kull 2004) though French colonial efforts to restrict mobility and force relocation for better tax provision decreased the surface area upon which to practice it (Jarosz 1993: 373).

French colonialists perceived *tavy* to be a destructive and inefficient land use activity (Jarosz 1993: 374). Similarly, Rio Tinto/QMM write, "slash-and-burn farming or 'tavy' and making charcoal, is the biggest factor in massive destruction of natural habitats... the company's environmental activities will lead to resulting benefit for this greatly endangered environment."[27] In reality, many other complex causes of deforestation exist in Madagascar and elsewhere (Lambin et al. 2001: 263; Angelsen and Kaimowitz 1999), such as tourism development, industry, logging, mining, or large-scale agriculture.[28] Various counter-arguments to the received wisdom about deforestation in Madagascar exist. For example, several figures of deforestation continue to uphold colonial accounts of annual landscape burning, which are highly unreliable and contradictory at best (Bertrand and Ratsimbarison 2004: 79). In turn, most international actors insist that, to date, Madagascar has lost up to 80% of its original forest cover (cf. Klein 2004: 14); however, this figure assumes that the island was originally covered in forest prior to the arrival of the first Malagasy inhabitants some 1500–2000 years ago; this view of forest loss has been challenged by Klein (2004) and Kull (2000) on the basis of a lack of scientific evidence to support the idea that the island was completely forest covered. Highland Madagascar was, for instance, probably covered in grasslands – not forest (Bourgeat 1972 in Klein 2004: 14). In turn, palynological evidence suggests that naturally occurring fires occurred long prior to human settlement and climatic factors played a key role in transforming landscapes in Madagascar (Burney 1997). In terms of

[27] Rio Tinto website, "Biodiversity – Background," http://www.riotintomadagascar.com/english/bioBackground.asp, Accessed 30.08.09.

[28] In an interview with an official for the "Ministry of Water and Forests" in March 2009 - a branch of the Malagasy government in control of both concessions and protected areas, local shifting cultivators (in Ampasy Nahampohana) were described as "les defraicheurs" (the "degraders") of the land.

contestations over forest use itself, Jarosz (1993) has shown that colonial policies of resource extraction (coffee cash cropping, logging concessions and "sustainable yield" ideologies) contributed to vast amounts of deforestation during the colonial period, and Kaufmann (2000) also points out how colonialists' introduction of invasive cochineal insects virtually obliterated a widely-used species of prickly-pear cactus (*raketa*) which was crucial to pastoralists for fencing, food and fodder for cattle (see also Middleton, this volume). Jarosz (1993) has also pinpointed the ideological bias of colonial forest policy, wherein Malagasy people were portrayed as "backwards," "barbaric" and rebellious (also see Bertrand and Ratsimbarison 2004: 80–81). Finally, Kull (2002, 2000) has described various local manifestations of forest protection often overlooked by conservation NGOs and the Malagasy state; for example, an ecosystem widely though to be degraded through unsustainable forest use, the tapia woodlands of highland Madagascar, have witnessed few changes over the past century (Kull 2002). In turn, local burning methods and uses of non-timber forest products offered by the ecosystems (e.g. fruit trees and silkworms) were employed to manage the forest and reflected sustainable natural resource use (ibid: 96). The remaining stands of tapia woodland are, consequently, uniquely adapted to fire, which is described as an important landscape management tool since human arrival on the island (ibid: 95–96).

These counter-narratives are important in that they add perspective to the deforestation claims made by Rio Tinto/QMM, which tend also to be ideologically-driven or re-invoke some of the myths described above. A widely-used Rio Tinto/QMM narrative follows that the littoral forests "would have gone anyway" due to local pressures on the environment; this argument serves as justification for the company's role in strip mining of 6,000 hectares of forest.[29] Manon Vincelette, former forest engineer for Conservation International and until recently head of the biodiversity programme at Rio Tinto/QMM,[30] was quoted as saying, "We are now saving forest that would otherwise go through slash and burn."[31] Echoing the

[29] The 6,000 hectare figure was quoted in the field and by various authors (cf. Parker 2004; Harbinson 2007); however, Rio Tinto/QMM suggest that mining would only involve the loss of 1217 ha ("Rio Tinto/QMM Biodiversity Offset Design", see references). This is clearly not the case as QMM (2007: 331) later claims that "rehabilitation of the mining site will involve an estimated 6,000 ha of land."

[30] Johnny Rabenantoandro is currently the head of the QMM biodiversity programme.

[31] BBC (2 November 2005) "Malagasy mine brings AIDs threat" by Olenka Frenkiel, Accessed online 20/08/09.

"victim" and "villain" debate detailed above, Vincelette et al. (2007: 4) state, "[t]hese villagers both endure and participate in a process of progressive deforestation and degradation of the environment." Such a statement is counter-balanced by Rio Tinto/QMM's claim to be bring economic growth to the region and contribute to poverty alleviation. Conversely, Virah-Sawmy (2009) has shown that conservationist discourses linking local pressures to littoral forest degradation – a narrative widely reiterated by Rio Tinto/QMM – are based on false assumptions about forest cover change over time. Using paleoecological evidence, she shows that the patchiness found in the littoral forest stems from complex climatic factors rather than long-term human interference alone, and that a majority of deforestation in these forests occurred the past twenty years, precisely coinciding with the exploration and infrastructural phases of the Rio Tinto / QMM project itself. Virah-Sawmy and Ebeling (2010: 1) mention that Rio Tinto's quantifications of "near – total forest loss on its mining sites in the absence of mining activities" are not correct.

By painting a picture of pre-existing degradation, Rio Tinto/QMM present a potentially "irrefutable truth"; namely, despite the impacts of the mine, Madagascar's rare ecosystems are on a collision-course with local populations. In QMM's 2001 SEIA (social and environmental impact assessment) the littoral forest is indeed described as already heavily degraded and fragmented due to local pressure (in Harbinson 2007: 46). Similarly, Daniel Lambert, director of QIT Madagascar Minerals, stated that "a majority of the proposed site where the deposit is has already been cleared or is highly degraded by the local community who need the wood for fuel" (Mines and Communities 2004).[32] This statement somewhat contradicts Rio Tinto/QMM's claim that the forests are extremely rich in terms of biodiversity (evidenced in the "biodiversity book" mentioned below).

Rio Tinto/QMM's "biodiversity book" (see photo 1 below) brings into question the company's claims to "transparency" within mining operations[33] as the narrative line in several of the contributing chapters focuses

[32] See MMSD (The Mining, Minerals and Sustainable Development Project) Final Report, pg. 262.

[33] It should be mentioned that while the EITI (Extractive Industries Transparency Initiative) focuses primarily on transparent disclosure of revenues and funds transfers, Rio Tinto/QMM have gone beyond this (outside of the EITI umbrella) to insist that mining operations are also transparent in the spheres of environment, sustainable development, human rights, and land access (see "Rio Tinto/QMM Social and Environmental Review," 2003).

Table 1. Terminology used in media disseminated by Rio Tinto/QMM to describe human environment (H-E) interactions in relation to Rio Tinto/QMM *vis a vis* local Malagasy people.[34]

| H-E Interactions: Rio Tinto/QMM | Relation to... | H-E Interactions: Malagasy people (land users) | Relation to... |
|---|---|---|---|
| "Impact(ed)" | Littoral forest; Biodiversity; Positive "Impacts" | "Degradation" | Littoral forest; Fort Dauphin locality; Land |
| "Unavoidable residual impacts" | Forests/biodiversity | "Heavily degraded" | Littoral forests |
| "Change" / "vicissitudes" | Environment; economy; culture | "Deforestation" | Littoral forest; Madagascar |
| "Remove(d)" | Littoral Forests | "(Over)exploited" | Natural resources; forests |
| "Sustainable" | Development | "Unsustainable" | Land use |
| "Training" | Local Fishing and agricultural practices | "Unproductive" | Agriculture |
| "Mitigation" | Socio-economic impacts of mining | "Poor/poverty" | Socio-economic status; World Bank indicators |
| "Improved" | Local Fishing and agricultural practices; access to markets | "Isolated" | Access to markets |
| "Restoration / Rehabilitation" | Forests (with eucalyptus) | "Destruction" | Littoral Forests / Ecosystems |
| "Harmonious integration" | Community Engagement | "Affected" by | Overall negative project impacts |
| "Appropriate" | Environmental Management | "Inefficient" | Fishing methods |
| "Opportunity" | Conservation; Economic development | "Pressure" | ["Increasing/Increased" human pressure] on ecosystem. |

[34] Various Rio Tinto/QMM documents and websites contain this language. Specifically, in Ganzhorn et al. (2007): "change", "vicissitudes" "destruction" "poverty" (11); "interference" (379); "exploitation" (59); "overexploitation" (380, 299); "massive deforestation" (380); "low efficiency" (295); "inefficient" (296); "opportunity" (7); "appropriate management" (306). Source also from Rio Tinto/QMM/QIT Madagascar Minerals Biodiversity

on local misuse of the environment and natural resource management rather than actual mining impacts. Acute impacts of mineral extraction on biodiversity or the wider ecosystem (e.g. via the creation of a dam to supply fresh water to the company), are never substantiated in the volume. As a result, with local pressures on land in the limelight,[35] several chapters tacitly lead to the conclusion that Rio Tinto/QMM are preserving biodiversity from local impacts on ecosystems rather than those of the mining project specifically.

The discursive tone and content of the chapters again underlines the deterioration of the forests by local people (an emphasis on human pressures). For example, while "deforestation" is an attribute of the local population, "change" or "vicissitudes" is used to describe Rio Tinto/QMM's own impact on forests (see above, table 1).

Table 1 details some of the language chosen by each contributing author in the biodiversity book to describe Rio Tinto's interactions with the environment *vis a vis* those of local Malagasy people; while the littoral forest will be "impacted" by mining operations, locals are leading to "degradation" – there is an important difference in the nuance. Similarly, "increasing pressure" on the natural environment (Sabatini et al. 2007: 299) and "unsustainable" (Rasolofoharivelo 2007: 63) land use practices have led to "massive deforestation" (Vincelette et al. 2007b: 379, 380) in the region.

While, on the one hand, the scientific publication appeals to an environmental aesthetic (rich, verdant and colourful biodiversity, e.g. the image used on the cover of the book), it also accentuates the apparent emptiness and highly degraded status of the forest. Contrarily, local Malagasy people's "maladapted materials and techniques" (Vincelette et al. 2007a) have led to "overexploitation" (Vincelette et al. 2007b: 380). The book convincingly positions Rio Tinto/QMM as an environmental NGO engaged in conservation and scientific research rather than a multinational company engaged in dredge mining. Because of the way in which the book is framed, several widely-used terms such as "rehabilitation" – which in practice *do* apply to Rio Tinto/QMM's own obligations to "reconstruct" ecosystems after mining with, for example,

Offsetting Strategy (e.g. "unavoidable adverse impacts"), and Rio Tinto/QMM website: http://www.riotintomadagascar.com/english/index.asp.

[35] Many of the chapters are also very "scientific" in nature and several focus on rehabilitation trials with fast growing species, inventories of plants, etc. Still the main *causes* of deforestation are often attributed to local (long-term) impacts rather than those of the mine.

eucalyptus plantations – appear to instead be directed at "restoring" an already highly "degraded" environment (caused by local Malagasy people). Rio Tinto/QMM write that, within the context of the Mandena littoral forest, "transfer to the communes will be decided… *when the population will be in a position to correctly manage the area* in a financially autonomous manner, supported by income-generating activities such as ecotourism, bee keeping and some others that are currently under development, in addition to revenue from the proposed Mandena plantations" of eucalyptus or acacia (QMM in WWF 2005: 66–67, emphasis added).

## *Conclusion*

This chapter discussed how multinational mining companies, Rio Tinto/QMM and Sherritt/Ambatovy, operating on two different large-scale mines in Madagascar, engage with biodiversity conservation and Corporate Social Responsibility (CSR) in Madagascar within the context of the emerging "green economy," markets for biodiversity offsets, mining-conservation partnerships, and a general trend towards "neoliberal conservation" in corporate best practice (see Igoe and Brockington 2007). Mining companies play an important role in anchoring regional sustainable development and conservation activities. Within this context, the corporate world is moving in the direction of "ethical capitalism" and forming partnerships with actors who hold "ethical" authority. As Barry (2004: 198) notes, ethical capitalism does not signify a new form of capitalism per se but rather a bundle of "techniques" that "have been developed to make the ethical or unethical conduct of businesses *explicit*" or visible and observable. Visibility is key to the formation of corporate ethics, and while some "problems and issues are made visible as ethical," others "are not" (Barry 2004: 196). Media is an important means in which mining companies display commitments to ethics and CSR (Kapelus 2002); on the internet, where websites, videos, and visually stunning photographs of rare biodiversity or smiling rural children at a school built by a mining company send numerous messages of best practice, immediate legitimization can be lent. Debord posited that mediation made up a central feature of late capitalism (in Igoe 2010). Just as social relations are mediated, so too are relationships between people are the environment (ibid). As much media produced by mining companies is cloaked in CSR language of accountability, dialogue and participation (Li 2009), it is important to consider how diverse stakeholders engage with both "nature" and mining projects through a set of significations – and in this sense, how the media used by Rio Tinto and Sherritt may create a "representational space" (Lefebvre 1991) wherein certain resources and/or species are abstracted and reabsorbed into an environmental crisis narrative.

For instance, Sherritt's use of the lemur (endemic to Madagascar) as a 'mascot' or flagship species in the company's conservation campaign feeds into broader narratives of the lemur's demise due to unsustainable farming practices and poaching. Sherritt writes:

> The Lemur Management Plan (LMP)" is "designed to protect lemurs living in the forests surrounding the Mine and ensure lemur viability. The LMP, in conjunction with a multifaceted 'Mine Area Forest Management Program', aims to decrease human pressure on the mine conservation forests by finding alternative activities that can fulfill the surrounding community's everyday resource needs.36

Here we can observe the inclusion/exclusion dynamics mentioned by Barry (2004); the selective invocation and omission of terms shapes the company's conservation rhetoric. As Escobar (1995: 39) points out, development creates "a space in which only certain things could be said and even imagined." In these spaces, much conservation discourse creates a "discursive blur" – a gap between rhetoric and reality that has a value all its own (Büscher and Dressler 2007). In the excerpt above, no mention is made of Sherritt's own impacts on lemur populations and habitats and the pressure this might create on surrounding environs; instead, attention is diverted to the local community's "resource needs."

Processes of legibility can also extend to what is deemed appropriate and inappropriate "use" of the environment (Luke 1997). Chairman of Rio Tinto, Sir Robert Wilson noted that "On the environmental side, it [mining] will use land, unnaturally change the landscape and cause some loss of biodiversity. But on the positive side, it may bring greater understanding and awareness of the environment and the harm caused by traditional practices, such as slash and burn (Wilson 2001: 4)."[37] Here the company not only makes what is important about "nature" legible to international audiences (biodiversity), it creates a "representational space" (Lefebvre 1991) that mediates a certain image of the forest as

---

[36] From http://ambatovyfiles.net/ebooks/ambatovy_pub2/HTML/#/2/, Accessed 03.09.12.

[37] See Wilson, Sir Robert. 2001. "How the Mining Industry Had to Dig Itself Out of a Hole" (The Guardian Observer, 8 July 2001). http://www.guardian.co.uk/society/2001/jul/08/business.theobserver8, Accessed 30.09.12.

degraded (purported an historically embedded result of local land us practices) and "empty" (see also Smith 1984). Whatmore (2002) has shown, for instance, how the concept of "wild" is produced through the material properties of particular animal species as well as those of specific spaces (e.g. "wilderness"). Media can thus represent at once 'empty' and 'biodiverse' space as well as 'wrong use' of the environment (see Luke 1997). Moreover, while biodiversity offsetting seems to promise "like for like" compensation for lost access, this chapter argued that locations and human interactions with the environment cannot be exchanged in this way in many parts of Madagascar, as no two places are the same if one considers the complex ways in which people are connected to a certain landscape or locality. Finally, the BBOP concedes that most compensation received is not "like for like," and that cash payments are more probable.

A critical analysis of Rio Tinto's "biodiversity book" and BBOP documents attempted to explore how mining's engagement in conservation may uphold the nature/culture dualism, which is problematic in Madagascar, where the human-nature dialectic is complex and involves multiple forms of valuation, stemming from ancestral and social relationships, perceptions of past, present and future, livelihood activities, and an overall sense of identity, as much of this volume speaks to. While both companies acknowledge human dependencies on nature to some degree, the process of reconciling the need for land both for mineral extraction and conservation (land set aside as offsets) precludes a truer understanding of local grievances, needs and voices. Protests such as those that occurred in January 2013 are testament to the gap between rhetoric espoused by Rio Tinto and local realities. This complicates the notion of providing "like for like" or "out of kind" compensation to people displaced from land. Biodiversity offsetting and mining's engagements in conservation appear to re-invoke the "nature-culture" divide (see Latour 1993) by conceptually separately social and environmental impacts, and severing people from the environment. In many documents, more attention is paid to research on biodiversity and conservation practice rather than local claims to land, which suggests a lack of nuanced understanding of local cultural and social dynamics, particularly pertaining to natural resource use. As Timothy Luke (1995: 21) questions, "Sustainable for whom?" While local populations have various methods of mobilizing land claims (and local forms of media), protesting against Rio Tinto is not just about dissenting against a mining company; these stakeholders also face the power and influence of historically embedded global environmental discourses and structures of sustainability that currently drive the global economy.

## *Bibliography*

Adams, W.M. and S.J. Jeanrenaud. 2008. *Transition to Sustainability: Towards a Humane and Diverse World.* Gland, Switzerland: IUCN, 1–108.

Adger, W.N.; T,A, Benjaminsen; K. Brown; H. Svarstad. 2001. Advancing a political ecology of global environmental discourses. *Development and Change* 32: 681–715.

ALT (Andrew Lees Trust) and PANOS. 2009. Madagascar voices of change: Oral testimony of the Antanosy people. London: Andrew Lees Trust & Panos.

Ambatovy Environmental Assessment (EA). 2006. Environmental Assessment Ambatovy Project, Summary. Submitted by Dynate Corporation of Canada on behalf of the Ambatovy Project (January 2006).

Angelsen, A. and D Kaimowitz. 1999. Rethinking the causes of deforestation: lessons from economic models. *World Bank Research Observer* 14(1): 73–98.

Anstee, S. 2007. Biodiversity offsets: a mining company perspective. BBOP Learning Network (Banbridge Island, PPT, June 2007). Available from: http://bbop.forest-trends.org/Presentations/BBOP5/June%2018/BBOP5_S_Anstee_%20What_are_Biod_Offsets.pdf [Accessed 18 February 2012].

Bakker, K. and G. Bridge. 2006. Material Worlds? Resource Geographies and the "Matter of Nature." *Progress in Human Geography* 30(10): 5–27.

Barry, A. 2004. "Ethical Capitalism," In: W. Larner and W. Walters (eds.), *Global Governmentality: Governing International Spaces.* London: Routledge.

Berkes, F.; J Colding and C Folke. 2000. Rediscovery of Traditional Ecological Knowledge as Adaptive Management. *Ecological Applications* 10(5): 1251–1262.

Bertrand, Alain and Ratsimbarison, Rivo. 2004. "Deforestation and Fires: The example of Madagascar," in Didier Babin (ed.) *Beyond Tropical Deforestation: From Tropical Deforestation to Forest Cover Dynamics and Forest Development.* Cirad-UNESCO: 79.

Borras, S.M. Jr.; R. Hall; I. Scoones; B. White and W. Wolford. 2011. Towards a better understanding of global land grabbing: an editorial introduction. *Journal of Peasant Studies* 38(2): 209–216.

Bourgeat, F. 1972. Sol sur solce ancient à Madagascar. *Mémoires* 57, ORSTROM, Paris.

Brockington, D., R. Duffy and J. Igoe. 2008. *Nature Unbound. Conservation, Capitalism and the Future of Protected Areas.* London: Earthscan.

Brockington, D. 2002. *Fortress Conservation: The Preservation of the Mkomazi Game. Reserve, Tanzania.* Bloomington: Indiana University Press.

Bryant, R.L. 1997. Beyond the impasse: the power of political ecology in Third World research. *Area* 29(1): 5–19.

Burney, D.A. 1997. "Theories and facts regarding Holocene environmental change before and after human colonization," In: S.M. Goodman and B.D. Patterson (eds.) *Natural and Human-induced Change in Madagascar.* Washington: Smithsonian Press, pp. 75–89.

Burnod, P.; Gingembre, M.; Andrianirina Ratsialonana, R. and Ratovoarinony,R. 2011. "From International Land Deals to Local Informal Agreements: Regulations of and Local Reactions to Agricultural Investments in Madagascar." Paper presented at the International Conference on *Global Land Grabbing* 6–8 April 2011, Brighton, UK.

Büscher, B. and Davidov, N. (eds.). 2013. *The Ecotourism-Extraction Nexus: Political Economies and Rural Realities of (un)Comfortable Bedfellows* (Routledge ISS Studies in Rural Livelihoods). London: Routledge.

Büscher, B.; S. Sullivan; K. Neves; J. Igoe, and D. Brockington. 2012. Towards a synthesized critique of neoliberal biodiversity conservation. *Capitalism Nature Socialism* 23(2): 4–30.

Büscher, Bram. 2010. Seeking 'telos' in the 'transfrontier'? Neoliberalism and the transcending of community conservation in Southern Africa. *Environment and Planning* A(42): 644–60.

Büscher, B. 2009. Struggles over Consensus, Anti-Politics and Marketing: Neoliberalism and Transfrontier Conservation and Development in Southern Africa. PhD Dissertation, VU University Amsterdam.

Büscher, B. and W. Dressler. 2007. Linking neoprotectionism and environmental governance: On the rapidly increasing tensions between actors in the environment-development nexus. *Conservation & Society* 5: 586–511.

Business and Biodiversity Offsets Programme (BBOP). 2009a and 2009b. *Biodiversity Offset Implementation Handbook*. BBOP, Washington, D.C.

Campbell, G. 2005. *An Economic History of Imperial Madagascar, 1750–1895: The Rise and Fall of an Island Empire.* Cambridge: Cambridge University Press.

Cardiff, S.G. and A. Andriamanalina. 2007. Contested spatial coincidence of conservation and mining efforts in Madagascar. *Madagascar Conservation & Development* 2(1): 28–34.

Castree, N. 2008. Neoliberalising nature: Processes, effects, and evaluations. *Environment and Planning A* (40): 153–173.

Castree, N. 1995. The nature of produced nature: Materiality and knowledge constructions in Marxism. *Antipode* 27(1): 12–48.

Cormier-Salem, M-C and T. Bassett. 2007. Introduction: Nature as local heritage in Africa: longstanding Concerns, new challenges. *Africa: The Journal of the International African Institute* 77(1): 1–17.

Corson, C. 2010. Shifting environmental governance in a neoliberal world: USAID for conservation. *Antipode* 42: 576–602.

Deacon, R.R. and P. Murphy. 1997. The structure of an environmental transaction: The Debt-For-Nature Swap. *Land Economics* 73(1): 1–24.

Demeritt, D. 2002. What is the "social construction of nature"? A typology and sympathetic critique. *Progress in Human Geography* 26: 766–789.

Duffy, R. 2006. Non-governmental organisations and governance states: The impact of transnational environmental management networks in Madagascar. *Environmental Politics* 15(5): 731–749.

Elmqvist T.E. et al. 2010. Biodiversity, ecosystems and ecosystem services. TEEB—The economics of ecosystems and biodiversity: the ecological and economic foundations. Available at: www.teebweb.org [Accessed 27 January 2012].

Escobar, A. 1995. *Encountering development, the making and unmaking of the Third World.* Princeton: Princeton University Press.

Evers, S. 2006. Expropriated from the hereafter: The fate of the landless in the Southern Highlands of Madagascar. *Journal of Peasant Studies* 33(3): 413–444.

Evers, S. and C. Seagle. 2012. Stealing the Sacred: Why "global heritage" discourse is perceived as a frontal attack on local heritage-making in Madagascar. *Madagascar Conservation & Development* 7(S2): 97–107.

Fairhead, J., M. Leach and I. Scoones (eds.). 2012. Special Issue: Green grabbing: a new appropriation of nature? *Journal of Peasant Studies* 39 (2).

Ferraro, P.J. 2002. The local costs of establishing protected areas in low-income nations: Ranomafana National Park, Madagascar. *Ecological Economics* 43(2–3): 261–275.

Fortwangler, C. 2007. Friends with money: Private sector support for national parks in the US Virgin Islands. *Conservation and Society* 5: 504–533.

Fox, L. 1990. *Hainteny: the traditional poetry of Madagascar.* Pennsylvania: Bucknell University Press.

Ganzhorn, J.U., S.M. Goodman and M. Vincelette (eds.). 2007. *Biodiversity, Ecology, and Conservation of Littoral Ecosystems in Southeastern Madagascar, Tolagnaro (Fort Dauphin).* Series editor Alfonso Alonso, SI/MAB Series #11. Smithsonian Institution, Washington DC, USA.

Genel, K. 2006. The question of biopower: Foucault and Agamben, Rethinking Marxism. A Journal of *Economics, Culture & Society* 18(1): 43–62.

Gezon, L.L. 1997. Political ecology and conflict in Ankarana, Madagascar. *Ethnology* 36(2): 85–100.

Hall, R. 2011. Land grabbing in Southern Africa: the many faces of the investor rush. *Review of African Political Economy* 38(128): 193–214.

Harbinson, R. 2007. "Development Recast? A Review of the Rio Tinto Ilmenite Mine in Southern Madagascar," PANOS London, research report for 'Friends of the Earth": pp. 1–70.

Harper, J. 2002. *Endangered Species: Health, Illness, and Death Among Madagascar's People of the Forest.* Carolina Academic Press.
Harvey, D. 2006. *Spaces of Global Capitalism: Towards a Theory of Uneven Geographic Development.* New York: Verso.
Harvey, D. 2005. *A Brief History of Neoliberalism.* Oxford University Press.
Hockley, N. and Andriamarovololona, M.M. 2007. An economic analysis of *Transfert de Gestion* in Madagascar: are donors looking for a free lunch? Ecoregional Initiatives Program/USAID Madagascar.
Horning, N.R. 2006. "Strong Support for Weak Performance: State and Donors' Mutual Dependence in Madagascar," IASCP Draft (Prepared for the International Study of Common Property's 11th Bi-Annual Meetings Ubud, Indonesia June 19–23, 2006).
Igoe, J. and D. Brockington. 2007. Neoliberal conservation: A brief introduction. *Conservation & Society* 5(4): 432–449.
Igoe, J. 2010. The spectacle of nature in the global economy of appearances: anthropological engagements with the spectacular mediations of transnational conservation. *Critique of Anthropology* 30(4).
Igoe, J. Neves, K, and Brockington, D. 2010. A spectacular tour around the historic bloc: Theorizing the current convergence of biodiversity conservation and capitalist expansion. *Antipode* 42: 486–512.
Ingold, T. 2007. *Lines: A Brief History.* London: Routledge.
Ingold, T. 2000. *Perception of the Environment: Essays in Livelihood, Dwelling and Skill.* London: Routledge.
Jarosz, L. 1993. Defining and explaining tropical deforestation: Shifting cultivation and population growth in colonial Madagascar (1896–1940). *Economic Geography* 69(4): 366–379.
Kapelus, P. 2002. Mining, Corporate Social Responsibility and the "community": The case of Rio Tinto, Richards Bay Minerals and the Mbonambi. *Journal of Business Ethics* 39: 275–296.
Kaufmann, J.C. 2000. Forest the numbers: the case of a Madagascar famine. *History in Africa* 27: 143–157.
Keane, W. 2006. "Subject and Objects," In: C. Tilley, W. Keane, S. Küchler, M. Rowlands & P. Spyer, *Handbook of Material Culture.* London: Sage Publications.
Keller, E. 2009. The danger of misunderstanding culture. *Madagascar Conservation & Development* 4(2): 82–85.
Keller, E. 2008. The banana plant and the moon: Conservation and the Malagasy ethos of life in Masoala, Madagascar. *American Ethnologist* 35(4): 650–664.
Klein, J. 2004. Fiddling while Madagascar burns: Deforestation discourses and highland history. *Norwegian Journal of Geography* 58(1): 11–22.
Kraemer, A. 2012. Whose forests, whose voices? Mining and community-based nature conservation in southeastern Madagascar. *Madagascar Conservation & Development* 7(2S): 87–96.
Kull, C. 2002. The "degraded" tapia woodlands of highland Madagascar: Rural economy, fire ecology, and forest conservation. *Journal of Cultural Geography* 19(2): 95–128.
Kull, C. 2000. Deforestation, erosion, and fire: Degradation myths in the environmental history of Madagascar. *Environment and History* 6: 423–450.
Lambin, E.F. et al. 2001. The causes of land-use and land-cover change: moving beyond the myths. *Global Environmental Change* 11: 261–269.
Latour, B. 1993. *We Have Never Been Modern.* Cambridge: Harvard University Press.
Lefebvre, H. 1991 [1974]. *The Production of Space* (transl. by Donald Nicholson-Smith). Oxford: Oxford University Press.
Li, F. 2009. Documenting accountability: Environmental Impact Assessment in a Peruvian mining project. *PoLAR: Political and Legal Anthropology Review* 32(2): 218–236.
Li, T. 2008. "The law of the project: government and "good governance" at the World Bank in Indonesia," In: F. von Benda-Beckmann, K. von Benda-Beckmann and J. Eckert, *Rules of Law and Laws of Ruling: On the Governance of Law.* Farnham/UK and Burlington/USA: Ashgate.

Littlewood, G. and T. Wells. 2000. The Global Mining Initiative. Melbourne Address to Mining (20 September 2000). Presented at the Minerals Council of Australia 2000 Minerals Industry Seminar. Available from: www.icmm.com/document/104 [Accessed 25 March 2012].

Liverman, D. 2004. Who governs, at what scale and at what price? Geography, environmental governance and the commodification of nature. *Annals of the Association of American Geographers* 94(4): 734–738.

Luke, T.W. 2005. Neither sustainable nor development: Reconsidering sustainability in development. *Sustainable Development* 13: 228–238.

Luke, T.W. 1997. "The (Un)Wise (Ab)Use of Nature: Environmentalism as Globalized Consumerism?" Paper presented at the annual meeting of the International Studies Association, Virginia Tech, March 18–22, 1997: 1–59.

Luke, T.W. 1995. "Sustainable development as a power/knowledge system: the problem of 'governmentality'" in Frank Fischer and Michael Black (eds), *Greening Environmental Policy: The Politics of a Sustainable Future*. Palgrave MacMillan.

Mail & Guardian. 2009. "Benchmarking the Responsible Approach." Supplement 3 (20 Oct 2009): 1.

McConnell, W. 2001. Putting people into the Earth system: How and why people and institutions matter beyond economy: People and trees in Madagascar. *Global Change Newsletter* 47: 20–22.

McDonald, K. 2010. The Devil is in the Biodiversity: Private sector 'engagement' and the restructuring of biodiversity conservation. *Antipode* 42: 513–550.

McNeilly, R J. "The Global Mining Initiative: Changing Expectations – Meeting Human Needs and Aspirations" Minerals Industry Seminar, Minerals Council of Australia, (June 7, 2000).

Merleau-Ponty, M. 2005[1945]. *Phenomenology of Perception*. London and New York: Routledge.

Mines and Communities. 2004. "Mining Giant Rio Tinto Defends Madagascar Project." (May 17 2004).

Myers, N. et al. 2000. Biodiversity hotspots for conservation priorities. *Nature* 403: 853–858.

Neimark, B. 2012. Green grabbing at the 'pharm' gate: rosy periwinkle production in southern Madagascar. *Journal of Peasant Studies* 39(2): 423–445.

Neves, K. and J. Igoe. 2012. Uneven development and accumulation by dispossession in nature conservation: Comparing recent trends in the Azores and Tanzania. *Tijdschrift voor economische en sociale geografie* 10(2): 164–179.

Peluso, N.L. and J.C. Ribot. 2003. A theory of access. *Rural Sociology* 68(2): 153–181.

Polanyi, K. (2001[1944]). *The great transformation: The political and economic origins of our time*. Boston: Beacon Press.

Prinsloo, L. "Green Accolade: Mining Group Lauded for Green Mining Efforts in Madagascar." *Mining Weekly* (Nov 6–12, 2009).

Rajak, D. 2008. Uplift and Empower: The Market, Morality and Corporate Responsibility on South Africa's Platinum Belt. *Research in Economic Anthropology* 28: 297–324.

Rasolofoharivelo, M.T. 2008. "Human Exploitation of Forest Resources in Mandena in 2000," In: J.U. Ganzhorn, S.M. Goodman, and M. Vincelette (eds.), *Biodiversity, Ecology and Conservation of Littoral Ecosystems in Southeastern Madagascar, Tolganaro (Fort Dauphin)*. Series editor Alfonso Alonso. SI/MAB Series #11, Smithsonian Institution, Washington DC, USA.

Richardson, T. and G. Weszkalnys. forthcoming. "Resource Materialities." Forthcoming in Anthropological Quarterly, Winter 2014.

Rio Tinto. 2010. QIT Madagascar Minerals SA: sustainable development report 2010. Available from: http://www.riotintomadagascar.com/pdf/SDR_2010.pdf [Accessed 19 February 2012]. Rio Tinto. 2007a. Rio Tinto QMM environment-biodiversity fact sheet. Available from: http://www.riotintomadagascar.com/english/pdfs/factsheets/QMM %20Fact%20Sheet%20 Biodiversity.pdf [Accessed 19 February 2012].

Roymans, N. et al. 2009. Landscape Biography as Research Strategy: The Case of the South Netherlands Project. *Landscape Research* 34(3): 337–359.

Sabatini, G., S. Salley and J-B Ramanamanjato. 2007. "A Review of the Spiny Lobster Fishery in the Tolagnaro (Fort-Dauphin) Region," In: J.U. Ganzhorn, S.M. Goodman and M. Vincelette (eds.). 2007. *Biodiversity, Ecology, and Conservation of Littoral Ecosystems in Southeastern Madagascar, Tolagnaro (Fort Dauphin)*. Series editor Alfonso Alonso, SI/MAB Series #11. Smithsonian Institution, Washington DC, USA.

Sarrasin, B. 2006. The mining industry and the regulatory framework in Madagascar: Some developmental and environmental issues. *Journal of Cleaner Production* 14: 388–396.

Scott, J. 1998. *Seeing Like a State: How Certain Schemes to Improve the Human Condition Have Failed*. New Haven: Yale University Press.

Scheper-Hughes, N. and M. Lock. 1987. The mindful body: A prolegomen to future work in Medical Anthropology. *Medical Anthropology Quarterly* 1(1): 6–41.

Seagle, C. 2012. Inverting the impacts: mining, conservation and sustainability claims near the Rio Tinto/QMM ilmenite mine in Fort Dauphin, Southeast Madagascar. *Journal of Peasant Studies* 39(2), 447–77.

Seagle, C. 2009. Biodiversity for whom? Local experiences and global strategies of land use and access near the Rio Tinto/QMM ilmenite mine in Fort Dauphin, Southeast Madagascar. Unpublished MA Thesis, VU University Amsterdam.

Shipton, P. 1994. Land and culture in tropical Africa: Soils, symbols, and the metaphysics of the mundane. *Annual Review of Anthropology* 23: 347–377.

Simsik, M.J. 2002. The political ecology of biodiversity conservation in the Malagasy Highlands. *GeoJournal* 58 (4): 232–242.

Sodikoff, G. 2005. Forced and forest labor regimes in colonial Madagascar, 1926–1936. *Ethnohistory* 52(2): 407–435.

Smith, N. (1984). *Uneven Development: Nature, Capital and the Production of Space*. Oxford & Cambridge, MA: Blackwell.

Sullivan, S. 2012. Banking Nature? The Spectacular Financialisation of Environmental Conservation. *Antipode* 45(1): 198–217.

Sullivan, S. 2010. 'The environmentality of 'Earth incorporated': on contemporary primitive accumulation and the financialisation of environmental conservation.' Paper presented at the conference A brief environmental history of neoliberalism, Lund University, Sweden (May 2010).

Swyngedouw, E. 2004. "Circulations and metabolisms: hybrid natures and cyborg cities." Paper presented to the research colloquium *Re-naturing urbanization*, held at the University of Oxford, 30 June.

Ten Kate, K., J. Bishop and R. Bayon. 2004. *Biodiversity offsets: Views, experience, and the business case*. Gland: IUCN, and London: Insight Investment.

Temple et al. 2012. Forecasting the path towards a Net Positive Impact on biodiversity for Rio Tinto QMM. IUCN and Rio Tinto Technical Series No.2. Gland: IUCN.

Tsing, A. 2000. Inside the Economy of Appearances. *Public Culture* 12(1): 115–144.

Van Dyke, F. 2008. *Conservation Biology: Foundations, Concepts, Applications* (2nd Edition). Springer.

Vincelette, M., L. Dean and J.U. Ganzhorn. 2007a. The QMM/Rio Tinto project history in Tolagnaro and its social and environmental concepts. In: J.U. Ganzhorn, S.M. Goodman and M. Vincelette, eds. *Biodiversity, ecology, and conservation of littoral ecosystems in Southeastern Madagascar, Tolagnaro (Fort Dauphin)*. Series editor Alfonso Alonso, SI/MAB Series #11. Washington, DC: Smithsonian Institution.

Vincelette, M. Rarivoson, C. and Anjara, B. 2007b. "Restoration Trials of the Anony River Estuary Banks by Dune Stabilization and Mangrove Tree Plantation," In: J.U. Ganzhorn, S.M. Goodman and M. Vincelette (eds.). 2007. *Biodiversity, Ecology, and Conservation of Littoral Ecosystems in Southeastern Madagascar, Tolagnaro (Fort Dauphin)*. Series editor Alfonso Alonso, SI/MAB Series #11. Smithsonian Institution, Washington DC, USA.

Virah-Sawmy, M. and J. Ebeling. 2010. The difficult road towards real-world engagement: conservation science and mining in Southern Madagascar. *Conservation Letters* 3(4): 1–2.

Virah-Sawmy, M. 2009. Ecosystem management in Madagascar during global change. *Conservation Letters* 2(4): 163–70.

Walker, P.A. 2005. Political ecology: where is the ecology? *Progress in Human Geography* 29(1): pp. 73–82.

Walsh, A. 2005. The obvious aspects of ecological underprivilege in Ankarana, Northern Madagascar. *American Anthropologist* 107(4): 654–65.

Welker, M.A. 2009. 'Corporate security begins in the community': Mining, the Corporate Social Responsibility industry, and environmental advocacy in Indonesia. *Cultural Anthropology* 24(1): 142–179.

West, P. 2006. Environmental Conservation and Mining: Strange Bedfellows in the Eastern Highlands of Papua New Guinea. *The Contemporary Pacific* 18(2): 295–313.

West, P., J. Igoe and D. Brockington. 2006. Parks and people: The social impacts of protected areas. *Annual Review of Anthropology* 35: 14.1–14.27.

Whatmore, S. 2002. *Hybrid Geographies: Natures, Cultures, Spaces.* London: Routledge.

WWF. 2005. An Update on the QMM Mining Project in the Anosy Region, Southeastern Madagascar. *WWF Review*: 1–76.

# HOW THE DAEWOO ATTEMPTED LAND ACQUISITION CONTRIBUTED TO MADAGASCAR'S POLITICAL CRISIS IN 2009

Venusia Vinciguerra[1]

*"Tsy amidy ny Tanindrazana"*
Ancestral land is not for sale

The acquisition or leasing of extended areas of land in developing countries by private firms is a growing phenomenon (Cotula et al. 2009: 4–5; Shepard and Anuradha 2009).[2] Some trigger factors for this are: food security issues tied to food prices, energy crops to respond to the growing demand for agrofuel, financial benefits, carbon markets and search for water (Grain 2009; Smaller and Mann 2009: 4–6). Countries with limited access to land and water constraints need to find alternatives to ensure food security for the domestic market. Moreover, one of the consequences of the rise in food prices is the increased value of arable land, making investments in the sector lucrative.[3]

Developing countries have a pressing need for investments; they need economic growth and foreign trade. Consequently foreign companies have a strong bargaining position while farmers, peasants and smallholders have little leverage (Braun and Meinzen-Dick 2009: 2). In addition, in many places locals tend to access land under customary arrangements and the lack of formal land titles worsens their bargaining position. As pointed out by the International Food Policy Research Institute (IFPRI), "land leases are often justified on the basis that the land being acquired by

[1] This research was carried out for *Senter for Interkulturell Kommunikasjon* (Norway) in the framework of partnership with the University of Toliara (Madagascar). My collaboration had a length of sixteen months, fourteen of which were spent in Madagascar (October 2008-December 2009).

[2] There are various ways to define this phenomenon. Academics prefer terms as "commercial pressure on land", "investment deals" or "large-scale land acquisitions"; while media and NGO's use less prudent terms like 'land grab" or "land rush". However, despite the feeling of "appropriation" that some of these deals recall, as pointed out by Etienne Leroy (*Journée foncier* 3.10.2009), these deals follow a legal path. Therefore, I will use 'large-scale land acquisition', as it better reflects the case analysed.

[3] "In 2007 alone, farmland prices jumped by 16 percent in Brazil, by 31 percent in Poland, and by 15 percent in the Midwestern United States"(Braun and Meinzen-Dick 2009: 1); Furthermore, several banks have set up agricultural investment funds, as for instance Deutsche Bank (Germany) or BlackRock (US), (Smaller and Mann 2009).

the foreign investor is 'unproductive' or 'underutilized' (Ibid.). However, land is often used for some activity, such as hunting, transhumant grazing, or wood collecting. Nevertheless, "these uses tend to be undervalued in official assessments because they are not marketable, but they can provide valuable livelihood sources to the poor" (Ibid.). IFPRI also highlights that further studies are needed to shed light on land deals. Documented information is scarce, resulting in a lack of transparency that limits the possibilities of the various local stakeholders.

## *A Political Crisis*

In January 2009, a political crisis erupted in Madagascar. In the capital Antananarivo, groups of town dwellers, led by the young former mayor Andry Rajoelina, gathered to protest against Marc Ravalomanana's government. There were three main causes for the demonstrations: (i) the purchase of a new presidential jet; (ii) the closure of Viva TV and radio broadcasting, owned by Rajoelina;[4] (iii) and South Korean Daewoo's plan to lease a large portion of land in the country (Pellerin 2009; Teyssier, Ramarojohn and Ratsialonana 2010: 5; Randrianja 2012: 15).

The 'Daewoo case' is the focus of this chapter. Given the importance of transparency in land deals, and the general lack of analysis of this type of transactions, the aim here is to outline the proposed Daewoo project in Madagascar; how the Malagasy government handled it; and in particular, how it is generally acknowledged as one of the causes of a *coup d'état* (Ploch 2009: 4; Randrianja 2012: 15).[5]

The academic literature on large-scale land acquisition is limited, while there is a considerable amount of information from the media. This chapter is based mostly on media sources, publically available documents and studies and information from key informants in Madagascar. The analysis was conducted at the same time of the events, thus semi-structured interviews were used to support and comprehend the dynamics reported by the media. Discourses, defined by Jorgensen & Phillips (2002: 1) as "a particular way of talking about and understanding the world (or an aspect of

[4] The official announcement at: http://www.madagate.com/politique/784-fermeture-de-la-television-viva-le-pouvoir-ravalomanana-a-visage-decouvert-.html (accessed 10/02/2009).

[5] Most of the international community defined the power shift as *coup d'état*, however in Madagascar itself people still fiercely debate whether the change of government was in fact a *coup d'état*.

the world)", were analysed to carry out this work. Discourse analysis is a widely used approach under the umbrella of "social constructionism" useful to explore social domains in different disciplines (Jorgensen & Phillips 2002: 4). The starting point is that reality is subject to change and people make choices to present their reality: people mould and edit their understanding, memories, stories and experiences into subjective discourses (Jorgensen & Phillips 2002: 145). Discourse analysis was a relevant framework for this research, as it highlights that different understandings of the world lead to different social actions. By looking at the discourses related to the Daewoo deal, it was possible to get an overview of the different reactions to it. Critics point out that if everything is subject to change, then there is no control in social life (Jorgensen & Phillips 2002: 6). Nevertheless, even though structures are socially created, they are hard to change, at least for dominated groups (Jorgensen & Phillips 2002: 6–7). This introduces the concept of power relations. According to discourse analysis, different discourses struggle for hegemony, understood by Gramsci as social consensus, in order to be at least temporarily unquestioned. The space for this struggle is identified as 'order of discourse': a social space in which discourses compete to infuse issues with meaning, to gain hegemony and power.[6] The order of discourse is a central concept here, as it defines the analytical space in which competing ideas related to access to land and foreign investment struggle to gain dominance. Furthermore, Dijk shows that discourses featuring people's beliefs are more persuasive, as they match their mental models. For instance when a social value is implicit in a message this can be taken for granted by the recipients and be understood without being openly stated (2001: 358). In this chapter various, often competing, discourses are represented concerning Daewoo's attempted land lease and the power struggle between Ravalomanana and Rajoelina. This approach revealed to be valuable in a study where the way stakeholders perceived and spoke about the land lease deal was central to understand their reaction to it.

Finding news and information sources in Madagascar during the unrests was at times challenging. Some of the media interrupted their broadcasts and distribution during the turmoil. It also should be noted

[6] Laclau and Mouffe's discourse theory operate with two concepts: 'discourse' which is the partial fixation of the meaning, and the 'field of discursivity', which encompasses all the other potential meanings. Jørgensen and Phillips add to this the concept of 'order of discourse' from Fairclough's critical discourse analysis, to delineate an analytical area in which not all the possible discourses, but only the competing ones are included (Jørgensen and Phillips 2002: 55–56).

that many of the sources used: newspapers, blogs and websites are not neutral (Randrianja 2012: 150–154). However, as they are often very explicit about it, it was possible to track down the different discourses they advocated (see note 22 for an example). On key topics, I followed discourses put forward by the media with different political views. This made it possible to follow how competing discourses were promoted. In addition, the professors at the University of Toliara were very helpful in explaining complex social mechanisms.[7] Finally, the rich ethnographic literature on Madagascar greatly assisted me in grasping relevant aspects of the social dynamics.

## *One Step Back*

Before delving into the Daewoo case it is useful to recap some background information of Marc Ravalomanana and Andry Rajoelina. This section is based on interpretation of events based on academic works, newspaper articles, interviews and other sources.

Marc Ravalomanana was born in 1949 into a dairy farmer family, apparently from the Merina middle class.[8] After a brief period in Europe, he obtained financing from the World Bank to establish his own dairy enterprise: TIKO. In 1999, Ravalomanana started an American-style campaign to become the capital's mayor; distributing t-shirts, TIKO products and offering free pop music concerts (Bornschein 2008). As mayor he introduced several innovations in the city, as well as several TIKO employees in the city's key posts (the "TIKO boys"), (Bornschein 2008). Ravalomanana decided to run for presidential election in 2001 as part of *Tiako I Madagasikara* (the 'I love Madagascar' party). Electoral results were contested by Ravalomana and culminated in an intense power struggle between him and the other candidate, Didier Ratsiraka (Fournet-Guérin

[7] Particularly Barthélémy Manjakahery, Lucien David and Pietro Lupo.

[8] Marcus (2004) uses the more radical term 'caste': the "Merina society is historically caste based, divided into four groups: *andriana* (nobles), *hova* (commoners), and *mainty* and *andevo* (slave groups). The caste system has diminished significantly in importance, however it continues to permeate key Merina political constituencies. According to Didier Galibert (2009: 451–2) Ravalomanana is from the *hova* intermediate caste, however his origins are largely speculated in Malagasy media. The aristocratic caste (Andriana) [of which Rajoelina is part] supported him in 2002 because he is Merina. However, when in power Tiko profited at the expense of many Merina-owned businesses, and a number of Andriana in positions of authority have suffered through his attempts to reconcile other groups". For a comprehensive study on the Merina grouping and social classes see: (Bloch 1971; Graeber 2007); and for the Betsileo living to the South of the Merina see: (Evers 2006).

2009). This lasted until May 6th 2002 when Ravalomanana was appointed president of Madagascar (Cornwell 2003: 6).[9] Ravalomanana's presidency is acknowledged for improving the country's infrastructure, the educational and health systems and despite criticism for the lack of progress in alleviating poverty, he was re-elected in December 2006 (IRIN 2006; Randrianja 2012: 241).[10]

Andry Nirina Rajoelina was born into the *andriana* (Merina) social class in 1974.[11] By 20, he had become a well-known disc jockey, and later set up a pop radio station (VIVA) as well as a successful advertising company (Injet). Rajoelina's profile began to rise with his increasing public appeal (Corbett 2009) and in 2007 he became the mayor of Antananarivo (Hogg 2009) through the movement *Tanora malaGasy Vonona* (Determined Malagasy Youth), and to some extent, by using the same successful campaign strategies as Ravalomanana. Rajoelina is nicknamed "TGV", as a reference to his movement's acronym, and to the French high-speed train: an expression of his energetic nature (Corbett 2009). Indeed, he offered a trendsetting alternative to the stuffy, older politicians (Hogg 2009). Rajoelina became one of the most determined critics of the president, setting an anti-government agenda (Corbett 2009).

The relationship between Rajoelina and the government further deteriorated at the end of 2008 when the Ravalomanana government closed Rajoelina's TV channel VIVA after the broadcasting of an interview with the former president, Didier Ratsiraka (Vinciguerra 2009). When Rajoelina's radio station was also shut down a month later, he convened a general strike in Antananarivo, accusing President Ravalomanana of corruption, authoritarianism and of selling ancestral land to foreigners (Ploch 2009). 40,000 people are reported to have attended the rally. On January 26th the demonstrations turned violent, and the public TV station, Ravalomanana's Madagascar Broadcasting System (MBS) and several department stores (including those of the president's TIKO and Magro) were looted and destroyed (CPJ 2009; Vinciguerra 2009). The following days the same happened in other cities of the country, with several deaths reported (Randrianja 2012: 16). On January 31st, Rajoelina proclaimed himself president of the High Authority of Transition, a parallel cabinet, requesting the President's immediate resignation (M.R. and E.R. 2009;

9 Marc Ravalomanana auto-proclaimed himself president on February 22nd 2002, without the recognition of the international community.

10 See also Randrianja (2012) chapter 10.

11 See note 8.

*Rajoelina demande* 2009; Randrianja 2012: 16). Ravalomanana refused, removing Rajoelina from office on February 3rd, when he also lost his bid to impeach the elected president through the High Constitutional Court (Ploch 2009). Rajoelina, who saw a reduced attendance at his rallies, called for another gathering to announce his government on February 7th.[12] Demonstrators marched toward the presidential palace, where the guards killed 28 people and injured several others (Vinciguerra 2009). The next weeks the two opponents called for different demonstrations, and attempted negotiations with the assistance of Council of Christian Churches (FFKM) failed (Vinciguerra 2009; Randrianja 2012: 104).[13] Rajoelina's ministers tried to take their posts, with limited success, but the situation escalated when on March 16th the Malagasy Army took over the Presidential palace.[14] Ravalomanana, who later declared that he was forced to hand power to the military on March 17th, left for an undisclosed location.[15] Vice Admiral Ramaroson was pressured to transfer his power to Rajoelina on March 18th, who suspended both chambers of Parliament the following day (Troops Out 2009; Madagascar Leader 2009; Ploch 2009).[16] The international community defined it a *coup d'état* (US Calls 2009) and several donors froze their cooperation with the country. Madagascar was also suspended from the African Union (AU) and from the Southern African Development Community (SADC), (McGreal 2009; SADC does not, cannot recognize 2009).[17]

## *The Daewoo Case*

On November 19th 2008, the Financial Times published an article describing a land leasing deal between Daewoo Logistic Corporation (hereafter Daewoo) and the government of Madagascar. The company was apparently planning to lease 1.3 million hectares to farm maize (1m hectares in

[12] As reported by Agence France-Presse the audience at the 4th of February rally was of 1,500 (Hundreds Protest 2009).

[13] More on the role of the Council of Christian Churches (FFKM) in Randrianja (2012) chapter 1 and 4.

[14] In the previous weeks the army witnessed a mutiny: the pro-opposition soldiers from CAPSAT (Army Corps of Personnel and Administrative and Technical Services), stormed the Army's headquarter and forced General Rasolomahandry to resign. (Civil War 2009; Madagascar President 2009).

[15] The location was Swaziland (Speech of His 2009).

[16] On legal aspects of the power shift: (Veum 2010; Randrianja 2012: 43–65).

[17] The European Community suspended their aid programme, due to lack of democratic progress (EU to keep 2010).

the western region), and palm oil (300,000 hectares in the eastern region) for a period of 99 years (Jung-a, Oliver and Burgis 2008). The Financial Times reported that Daewoo signed a memorandum of understanding in May 2008, and a final contract in July 2008, as stated by Hong Jong-wan, manager at Daewoo. The "bulk of the harvest" was to be shipped back to South Korea to boost the country's food security. It seemed that Daewoo would have to pay nothing for the lease, however Madagascar would benefit with employment opportunities and infrastructure (Jung-a, Oliver and Burgis 2008).

### *How was Daewoo Planning to Implement the Investment?*

Despite the substantial amount of articles, it is not simple to acquire a precise idea of what Daewoo was going to do in Madagascar, and how they were planning to do it. Most news was an echo of the Financial Times article. However, on November 21st 2008, *Le Monde* reported that Panja Ramanoelina, Minister of Agriculture, Farming and Fishery, and Marius Ratolojanahary, from the Ministry of Land Management and Reform, insisted that Daewoo's project was still in the evaluation phase (Tuquoi, Messmer and Hervieu 2008). While President Ravalomanana only one week earlier denied any knowledge of the existence of the project, the local press reported that land requests were deposited at the offices of Toamasina and Brickaville (east Madagascar), and Besalamby (west Madagascar), (*Investissement Agricole* 2008). By the end of December, the government had again denied any advanced phase of the project, but a web source revealed that from October 6th 2008, portions of land were already leased with the authorization of decentralised state authorities (*Foncier la Presse* 2008). The position of the government was however backed up by Daewoo, who accused the Financial Times' article of being riddled with errors (Navalona 2009; Daewoo Logistic Corporation 2008).

The government never shifted its position from saying that Daewoo filed an investment request that was in its evaluation phase as normal procedure. However, the deal with Daewoo became one of the "battle horses" of Rajoelina's protest against the regime (*Madagascar: l'Affaire* 2009; Randrianja 2012: 15, 197). Rajoelina declared that Ravalomanana was selling the country to foreigners, while according to him in Malagasy tradition, land cannot be sold or leased to non-Malagasy (Carayol 2009; *Madagascar: l'Affaire* 2009). The same day he took power (March 18th 2009), Rajoelina cancelled the deal with Daewoo as "in the constitution it is stipulated that Madagascar's land is neither for sale nor for rent, so the agreement with Daewoo is cancelled" (Madagascar Leaders 2009). "We are not against the idea of working with investors, but if we want to sell or rent out land, we have to change the constitution, you have to consult the people" (Madagascar Leaders 2009). Daewoo declared that they had not received official communication of the investment cancellation (Ramaro and Sguazzin 2009), however, on April 10th the news website Orange Madagascar, published a press release from the company notifying the end of Daewoo's project in Madagascar (*Communiqué de Presse* 2009).

The company Madagascar Future Enterprise (MFE) managed the deal from the capital, Antananarivo. According to sources who wish to remain anonymous, the company organized survey flights in October 2008, both with a helicopter rented from a company called Madhelico, and a private plane provided by a company called Aeromarine.[18] The helicopter was used to scrutinise the eastern part of the country: from Antalaha (north east) to Mananjary (central east). The plane was employed more than once, to take aerial photos for agricultural evaluations, and to transport "important people from the South Korean company to the sites".[19] According to the sources, the South Korean financial director of MFE first covered the east coast: from Toamasina to Anstiranana, passing over Antalaha and Samabava and then from Toamasina to the south until the area around Manakara. Then the plane was used to fly west of Antananrivo, over the Melaky region, where a test farm was implemented. One of the informants claimed to have seen pictures of corn cultivations in Melaky in autumn 2008, and had a conversation with a Daewoo executive about the test plantation. This person stated that South Koreans as well as Malagasy people from Daewoo went to the Melaky region, to check on the project. The informants added that negotiations in the area were done with the chief of region in Maintirano (Melaky), and that the field workers were local Malagasy. At the end of December, the media reported that land from the east side of the country had already been leased on October 6th, and that the papers had been signed by the chief of region and district together with authorities from the decentralized service and the state (*Fonicer. La Presse* 2008). Another article reported that after the political

[18] The people disclosing information work for three different companies. The information was crosschecked. The informants want to keep their identities anonymous so as not to damage the companies they work for, of which Madagascar Future Enterprise was a client. Furthermore, given the unstable political situation in Madagascar at the moment, the informants also prefer to stay anonymous.

[19] Those are the exact words in English of one of the informants interviewed.

unrests and constant press attention, Daewoo's project changed name from "Madagascar Future Enterprise" to "Madagascar Tsako SARLU" but kept its office in Antananrivo (Land Grabbing 2009). Under this new name, according to the article, Daewoo made plans to achieve the lease of 10,000 ha in the SAVA region to grow maize. The supposedly closed office of MFE in Antananarivo, although had no name plate, was stil open as late as august 18th 2009.[20]

Information on when and where Daewoo through MFE was planning to lease land differs. However, most discourses shared the following notions: (i) the company had plans to lease a large area of land in the country, and (ii) the terms of this lease were not clear.

### *How did the Government Manage Daewoo's Application?*

#### *Legal Basis*

Statute n. 2007-036 of 14 January 2008 regulates land investments in Madagascar. Foreign investors (with an investor visa) can buy Malagasy land under the following conditions:

1. the land must be used for professional exploitation. The government can withdraw the title if this condition is not respected;
2. the foreign investor needs to apply for the "authorization for land acquisition" to the EDBM (Economic Development Board Madagascar): a formal approval to legally purchase land;
3. the foreign investor needs to submit a business plan to the EDBM, to illustrate in detail the proposed investment (equal to or more than 500, 000 US dollars), (Malagasy Land 2008).

Moreover, to access this procedure the foreign investor needs to act through a Malagasy legal entity.

According to Ravalomana's government, this procedure was being followed. No land had been sold or leased to Daewoo or to its Malagasy legal entity, and a request was deposed to the EDBM, following procedures. The minister of Estate and Land Management, Marius Ratolojanahary, specified that the evaluation of Daewoo's request pertained his ministry, the ministry of Agriculture, Livestock and Fishery and the EDBM.[21]

---

[20] Behind some vacant rooms, some employees worked silently. No one agreed to release an interview about MFE's land leasing projects.

[21] My translation of : Ministère des domaines et de l'aménagement du terroire and of Ministère de l'Agriculture, de l'Elevage et de la Pêche. (Felana 2009).

An inter-ministerial committee was going to be established to evaluate Daewoo's business plan; however, the final decision was to be taken by the council of the ministers. The minister added that the chiefs of regions play the role of facilitators in this type of investment and citizens can express their opinions. He also specified that no land expropriations were expected (Felana 2009).

The representatives of Daewoo backed up the position of the government. However, in January 2009, Shyn Dong Hyun, the financial director of the company, declared that the project was suspended due to the political situation in Madagascar and to the financial crisis (Ramiakajato and Ravelomanantsoa 2009: 16).

### *How the Deal was Handled, According to the Press*

In mid-November, the press signalled that despite the declarations of the ministers, land requests were already deposited in Toamasina and Brickaville (east of Madagascar) and Besalampy (western part of the country) in October (*Investissement Agricole* 2008). According to an article, Madagascar Future Enterprise filed land requests at the estate office of Toamasina (file n° 24 294 TAM), Brickaville (n° 24 293 BRICKA) Vatomandry and Melaky. The tabloids considered this as a proof that Daewoo's request was no longer in the evaluation phase (Maka 2008).[22] Moreover, the *Fokontany* and the municipalities were not informed, therefore no one could express an opinion (Maka 2008).[23] The newspapers reported that the officials at the regional office of Toamasina refused to answer any questions on the matter; while an "informed but anonymous source" said the chief of region had already approved the land requests (Maka 2008). The legal journal, *La Revue de MCI*, presented a legal analysis of the Daewoo case under the assumption that what the government had declared was genuine (Ramiakajato and Ravelomanantsoa 2009). The investigation showed the procedure imposed was correct, as no legal provision was violated. Nonetheless, the journal denoted the possibility of a verbal contract between the parties, a suspicion reinforced by the lack of transparency (Ramiakajato and Ravelomanantsoa 2009: 16–17).

---

[22] It has to be underlined that some reporters and newspapers were openly pro Ravalomanana or pro Rajoelina. For example MAKA from La Vérité, was in this second group, and later became a significant figure in the HAT government, see also chapter 6 in (Randrianja 2012) on this topic.

[23] The *Fokontany* is the smallest administrative unit in Madagascar.

Once more, information from the press, from the government and from the opposition did not always correspond. The whole case was not clear and transparent to the public, contributing to the civil and political instability that occurred.

### *How did Malagasy React?*

The reactions regarding the land lease requested by Daewoo have been various. Some discourses stressed people's concerns and fears, while others were open and optimistic about the investment. To try to gauge the different opinions discourses were analysed through newspaper articles and websites, as well as readers' comments appearing in articles and blogs. The fact that media in Madagascar are mostly politically aligned and that only a limited amount of Malagasy have access to the Internet or other media was also taken into consideration (see also note 22). Many Malagasy are illiterate or do not speak French and newspapers are not distributed everywhere or arrive with some days' delay. Despite a large number of TV channels accessible in the capital, most of the island receives one or two channels, if any (Randrianja 2012: 149).[24] The ethnographic literature was of help in trying to fill the gaps in discourses.

Global Voices Online reported that "reactions to news of the land deal were heated and diverse in the Malagasy blogosphere"; the Malagasy website, Sobika wrote about the deal quickly after the Financial Times asking their readers to comment on it (Rakotomalala 2009). Not long after, more than 100 comments were posted. Two main discourses were identified: a "traditional" discourse where concerns were expressed about the possibility of ceding land to foreigners; and an economic discourse, which focussed on the benefits for Madagascar (Rakotomalala 2009). Müller and Evers (2007) highlighted that "just as in many other African countries, land in Madagascar holds great emotional value, thus giving land a deeper meaning than that of purely being economic property". People leaving comments on websites and blogs recurrently used the word *sacred* when referring to land. Expressions like *mpivarotra tanindrazana* respectively meaning "sellers of homeland" and "ancestors land", are frequently found to communicate that "land is sacred for all Malagasy" or that "real Malagasy would never give away their land to foreigners" as "land is the sacred place of the ancestors" and "selling land is an irreversible treason of the sacred

[24] See also chapter 6 in (Randrianja 2012).

for Malagasy".[25] On January 27th a group of citizens attacked the Japanese company Daiho Corporation throwing stones and screaming "you're land thieves, you want our land!" because they had mistaken the name Daiho for Daewoo.[26]

Following the protests in front of the Malagasy embassy in Paris, most of the Malagasy opposition parties worked together to present an official complaint against Daewoo to the President of the *Conseil d'État* (*L'Opposition Porte* 2008). In the same period an association was founded in Paris, the *Collectif pour la défense des terres Malgaches* with the goal of "contributing to defend Malagasy land and to sustain Malagasy citizens and peasants in their fight for to defend their land and their development".[27] This group planned to defend Malagasy land by spreading information about large land deals and by creating a network between countries "victim of similar land grabs" (*Terres Malgaches*). The collective prepared a letter of complaint to Daewoo's CEO, to ask to "suspend the ongoing actions and land survey", to ask for transparency and to respect the farmers' rights to use land (*Terres Malgaches*). For the group, the elements making the deal unacceptable were: the size of the land which would necessarily involve sacred ancestral land; the dangers for the farmers dependant on this land; the fact that the procedures followed by Daewoo were given priority over those of small farmers; the fact that the products of the cultivations would be exported; and finally the danger that the cultivation would spoil the terrain, destroying the biodiversity.

According to the international press, there is no doubt that Daewoo's investment plan in Madagascar played a role in the country's power shift. It is thus of great interest to analyse what aspects of Malagasy society were used in discourses and specifically contributed to make people feel menaced by the deal.

## *Why did many Malagasy React so Strongly?*[28]

Certainly the reasons for Malagasy to oppose the Daewoo deal were diverse and nuanced. The historical background might have played a role,

[25] My translation from comments appeared on: (Démenti de Daewoo 2009) See also: (Razafiarijaona 2009).

[26] My translation from: (Andriamanana 2009).

[27] My translations from www.terresmalgaches.info (accessed 10.10.2009).

[28] Not all the Malagasy were in the position to participate in the debate about the land lease. However, although the Malagasy generally are quite reserved and indirect in expressing their opinions, when it comes to land issues many seem to be willing to protest openly.

as well as people's culture, and general political interests. A systematization of the background and causes that lead to such opposition is needed. To do so, the following section is organised in six parts outlining: Malagasy's attachment to land; their fear of a return to colonialism; the general problem of land management; the sense of distrust of the Ravalomanana's administration; as well as the economic and political reasons to oppose the deal.

### *Strong Attachment to Land*

"Earth is the first wife of the Creator. She amasses the living people and wraps the dead ones" (Jaovelo-Dzao 1996: 15). Land for Malagasy is a place to live as well as the link between the living and the ancestors (Fournet-Guérin 2009). The word "*tany* or land has about 11 equivalents in English varying from land to earth, country, soil, field, place or environment, world, native land, estate, ground or life" (Aliderson 2004). According to Fournet-Guérin (2009), the notion of *tanindrazana* is an element of national unity, also expressed by the start of the national anthem: "*Ry tanindrazanay malala ô!*", "oh, beloved ancestor's land". Two types of land can be distinguished in Madagascar "that of the ancestors which has 'always' belonged to the family and contains the tomb, and that which was purchased later to increase agricultural production. In theory the first can only be transferred to the heirs, whereas the latter can also be sold to other parties if necessary" (Bloch 1971: 127–130; Van den Heuvel and Evers 2007). "The importance, in all Malagasy societies, of having a strong link to one's *tanindrazana*, can be hardly exaggerated", notes Keller (2008: 652), adding that the importance of this element is shared throughout Madagascar. The ancestors are the authorities that need to be addressed when asking for fortification, protection and approval. They are always evoked together with the creator, in both family and political speeches (Aliderson 2004; Rakotoarivony and Ratrimoarivony 2006). The ancestors are interrelated with the idea of land: Malagasy take a handful of the native-land soil when going abroad for a long time as an "embodiment of the support and blessing from the whole family dead or alive and a protection from the ancestors as well as a hint of their presence" (Aliderson

When I refer to Malagasy here, I refer to those who were able to access the debate about Daewoo's investment and/or those participating in the protests. The point here is to identify what kind of discourses where used in the debate, and why some of them functioned as trigger in the protests.

2004). For the similar reasons some bury the umbilical cord of newborns in the land of their origin (Rakotoarivony and Ratrimoarivony 2006: 4), the same land that will host the body after life, an affiliation that stands even if the person has lived far away from the family tomb. A study on migration to Antananarivo shows that "even the second generation of migrants born in Antananarivo wish to be buried in the family tomb in the village of origin" (Van den Heuvel and Evers 2007). Those are more than simple signs of nostalgia, they are signs of a feeling of belonging that Malagasy have to their *tanindrazana* (Rakotoarivony and Ratrimoarivony 2006: 4). In everyday life, land became a mythical element interrelated to both the past (creator and ancestors) present and the future: owning land is symbol of wealth as well as of being alive. This concept is well outlined by Keller: people not only need land for their present survival, but they need it in order for their descendants to create *tanindrazana* (2008: 656). People without *tanindrazana*, cannot receive ancestral blessing, and will possibly "stop being a full person" (Keller 2008: 659; Bloch 1994). In her study in the southern highlands of Madagascar, Evers (2006: 417) points out that those who are unable to claim ancestral homeland "do not have history", which is a cause of marginalisation.[29] Tombs and ancestors are a way to establish and perpetuate control over land, and labourers lacking *tanindrazana*, cannot claim property rights (Ibid.). Accordingly, owning land gives the assurance of having a say in any communitarian decision and in social conflict (Aliderson 2004).

A deal involving 1,5 millions of hectares of land naturally caught people's attention, and the concept of *tanindrazana* was central in discourses used to oppose the Daewoo investment. We have seen that land is considered ancestral when tombs are on it; nevertheless in some discourses this concept was used more generally, even to designate Madagascar in general. *Tanindrazana* was to a certain extent manipulated in discourse to trigger emotions. The ancestors are believed to have an influence in people's lives, and problems such as misfortune and illness, are believed to be sign of ancestral discontent (Astuti and Harris 2008: 715–718). People feared that leasing part of their ancestral land to foreigners would show disrespect towards the ancestors, with negative consequences. We should not think that those matters involve only "superstitious people". The notions of ancestors and land are so interrelated that disrespecting them

[29] A marginalization that in certain situations is equivalent to being put in the slave, impure, category. Thus, *tanindrazana* becomes important also to change this and overcome the slave status. (Evers 2006; Keller 2008; Graeber 2007).

would be for many Malagasy like betraying their understanding of social identity. Discourses used by Rajoelina and his supporters, but also by some media and exponents of the Malagasy diaspora, made considerable use of concepts related to the ancestors and ancestral land. In the same way such notions were powerful tools in Rajoelina discourses, awakening fears related to the colonial period.

### *Fear of Neo-Colonialism*

"Land is here a sensitive topic [...]. People are not willing to live once again the ordeal of colonisation, when their fields were confiscated for revenue cultivations which brought great advantages to the colons but no nourishment to Malagasy" (Carayol 2009). This sentence expresses the feelings of those fearing a neo-colonial era in the country. Madagascar regained independence in 1960 after French domination since 1895 (it became a French colony in 1896) and a revolt in 1947, which caused thousands of deaths and is still a powerful memory for many Malagasy.[30] Some thought that leasing or selling land to non-Malagasy would be an act of colonialism "made by Daewoo". It would be like neglecting the bloodshed of the martyrs who fought against the colonial oppression (Andriamananoro 2009). Western discourses denouncing the deal as an act of neo-colonialism, were also numerous in the international press (Jung-a, Oliver and Burgis 2008; C'est du Colonialisme 2009; Ryall and Pflanz 2009); and included Jacques Diouf, the head of the FAO, who welcomed long term agricultural investments, but worried about the fact that many of those deals looked more profitable for the investors than for the hosting countries (Ryall and Pflanz 2009). Some argued that the preoccupation with denouncing Asian investors for their neo-colonialism was in reality concealing Europe's own interests in Africa (Rakotomalala 2008).

Another historical episode, preceding colonialism, shows Malagasy's resistance to ceding land to foreigners. Prior to the French domination, various kingdoms ruled different areas of Madagascar. In the beginning of the 19th century, the Merina rulers extended their domain from the highlands to much of the island. King Radama II came to power in 1861 showing clemency towards foreigners, who had been persecuted by his

30 For more information about the colonial period in Madagascar, and the land politics of the period see: (Ellis 1985). For the relevance of the 1947 Malagasy Uprising see: (Hanson 2007).

mother, Queen Ranavalona (1828–1861). He reopened the Malagasy kingdom to foreigners, and signed documents according the settlers the right to exploit land (Dahl 2008: 82–95). Despite the clear discontent of the population, the king refused to renege, persisting with his openness toward foreigners. In 1863 his close counsellors strangled him. Giving land to foreigners: an act of treason to be punished exemplarily, to the extent that his name was expunged from the official list of the country's royals (Délivré 1974: 373). The successor to the throne, Queen Rasoaherina, put back into place the interdiction to lease land to foreigners, a decision that none of the following rulers, until Marc Ravalomanana, tried to change (Ramambazafy 2009). According to journalist and agricultural consultant Chido Makunike (2009), "paying attention to the recent political history of Africa should have warned all parties concerned in this deal that strong opposition to it was not only predictable but almost guaranteed". There are two main reasons for this: (i) the general association, following the colonial times, made between large plantations, usurpations and plunder of resources; (ii) many African countries have not yet found an adequate land tenure system (Makunike 2009; Smith 2003; Marc Ravalomanana 2009). Madagascar is no exception, and many feared for the future of those working the fields without official land titles.

### *Problems with Land Management*

Lack of land security is a widespread problem in Africa (De Shutter 2009: 7). Customary law permits users to access land without legal titles while legal practices can be source of uncertainty as land users have limited access to legal instruments. The absence of legal titles gives land users little chance to be fairly compensated if land is expropriated, or leased (Ibid.). Activities like grazing, or beekeeping, are often neglected by compensation, despite their importance for people's livelihood and culture, as anthropologist Marco Bassi of Oxford University points out (Bassi in Mackenzie 2008 and Cotula 2007: 76). In addition, fallowing practices used by nomadic cultivators, can make land left fallow appear unused to potential investors (FAO 2009: 2; De Shutter 2009: 12; Mackenzie 2008). Nevertheless, as pointed out by the International Land Coalition "seemingly unoccupied land is probably used for at least part of the year by someone" (Mackenzie 2008). In addition, seasonal workers or people renting and not owning the fields formally could be left out of the negotiations, losing the fields they cultivate without holding rights to compensation.[31] Finally, IFPR highlights that "Many tropical soils are unsuited for intensive cultivation (one reason for long-fallow cultivation cycles in many tropical areas that are considered 'unused'), or there is insufficient water for intensive cultivation" (Braun and Meinzen-Dick 2009: 3). The danger is that after intensive exploitation land and water are left exhausted, instead of enhanced for further use (Mackenzie 2008).

In Madagascar both the "control and maintenance of land access has been a complex policy issue for years" (Pronk and Evers 2007). The French Civil Code introduced during colonization, brought with it rules contradicting the traditional customs of land management. Malagasy customary law has roots that are deeply linked to the respect of the ancestors (Razafiarijaona 2009: 65,77,90; Evers 2006; Keller 2008; Graeber 2007; Bloch 1971) and land is perceived as a common resource to the extent that private ownership of *tanindrazana* is considered inappropriate, as it is traditionally protected by kinship relations.[32] At the same time, there is a growing individualism that sees land as a commodity (Teyssier, Raharison and Ravelomanatsoa 2007: 36). Over the years Malagasy governments have faced challenges due to the contradictions between traditional land tenure and legal rules, most of which remained unsolved, resulting in a sort of legal pluralism that boosts land access insecurity (Pronk and Evers 2007). As Pronk and Evers (2007) report in their study on the complexity of land access in southeast Madagascar, only about the 10% of land was registered in 2005, while at the time of writing about 70% of the fields in rural areas belonged to communities according to customary rules, and not to individuals (Razafiarijaona 2009: 78).

In 2005 Ravalomanana's government, pressured by the International Monetary Fund, started the National Land Tenure Program to increase the number of land titles (Madagascar Action Plan 2007–12 and Evers 2006). In the previous century only 330,000 titles were issued, while the Madagascar Action Plan set the goal of 75% of the farmers obtaining land titles or certificates by 2012 (Madagascar Action Plan 2007–12: 64). In 2008 48,58% of farmers still had no knowledge about the procedures to follow (*Foncier: l'Affaire* 2008). The government also introduced another land category:

[31] Renting here refers to people receiving the right to use of land in exchange for money or part of their crop.

[32] Also in the opinion of Professor P. Lupo (personal communication in Toliara) according to Malagasy custom land belongs to the collective, which manages it through the *Fokolona*. An opinion confirmed in: (Razafiarijaona 2009) See also: (Pronk and Evers 2007).

the non-titled-property, limiting the presumption of State property of non-titled territories (Razafiarijaona 2009: 70). Nevertheless, obtaining a title can be still too costly for poor farmers or too complicated for illiterate people.[33] Not having a land title might scare those conscious that they risk losing their land when their ownership is not confirmed in a title (Pronk and Evers 2007); while others believe that to follow customary rules is enough as the land belongs to their ancestors, furthermore this will avoid the costs of registration and taxes (Evers 2002; Müller and Evers 2007; Razafiarijaona 2009).

Given the lack of land titles and the amount of land unofficially occupied, it seemed questionable when the government declared that the land requested by Daewoo was unoccupied and that no farmer would be displaced. Particularly considering that, according to the same government representatives, Daewoo was still at an early phase of prospection (Navalona 2009; Felana 2009).

Several researchers have pointed out that uncertainty around access to land is a growing source of conflicts (Teyssier, Raharison and Ravelomanatsoa 2007; Evers et al. 2006; Müller and Evers 2007). Pronk and Evers (2007) signal that the statement of the United Nations Development Program (UNDP): "access to land is one of the main determinants of poverty", applies to "large parts of the Malagasy population". In the same article they report: "In an interview, a representative of the Food and Agricultural Organisation (FAO) argued that in the whole of Madagascar there is almost no fertile land to be found that has not already been cultivated. Since the population of Madagascar is rapidly increasing" (Pronk and Evers 2007).

Another problem is the unbalanced access to land, with consequential tensions between different social groups. Governmental employees have reportedly used their leverage to access land. In Menabe, one of the areas under Daewoo interest, people with access to local power commonly managed to settle land disputes to their advantage Teyssier, Raharison and Ravelomanatsoa 2007: 39). For instance, in 2004 the municipality of Ankilivalo (Menabe), prevented the intervention of the state topography teams in the area, as they did not trust their evaluations (Ibid.: 15). This example confirms that Malagasy farmers live with a feeling of uncertainty when it comes to land (*Foncier: l'Affaire* 2008), and shows that farmers

[33] The costs are equivalent to $500, and the procedure lasts an average of six years (Teyssier, Ramarojohn and Ratsialonana 2010: 10).

develop a sense of distrust toward the authorities, a constant fear of power abuses.[34]

*Distrust*

"Laws were made that the stronger might not in all things have his own way" (Ovid, in Cotula 2007). A population accustomed to corruption is not new to feelings of suspicion toward the authorities and powerful people. Even those in charge of fighting corruption were looked at with a certain amount of distrust.[35] Malagasy people had a lot of hope placed in Ravalomanana. Nevertheless in his presidency he largely mixed state affairs with private business: several TIKO employees worked in public office, and through the removal of taxes, control of market prices, etc., TIKO's profits quadrupled since 2002 (Pellerin 2009); "The presidential airplane is registered under TIKO's name. Asa Lalana Malagasy, a TIKO sub-corporation, has become the de facto public roads company [...]" (Marcus 2004).[36] According to Christine Razanamahasoa, justice minister of the HAT, Ravalomanana "mixed public interests with his personal interests", and soon after the power shift he was sentenced for "alleged abuse of office" (Madagascar Sentences 2009). Many think he was governing the country as if it was his personal enterprise instead of a constitutional republic (Adriamananoro 2009).

Lack of transparency boosted sentiment of distrust. In this direction moved a particular kind of discourse used by professionals and educated Malagasy, the lack of transparency favours corruption and "fuelled the suspicions that the businessman-president Ravalomanana manages the country like it is private property. Land included" (Carayol 2009).[37]

Farm Radio, an organisation helping and informing farmers in developing countries, made a list of questions one should ask when a government is dealing with farmland: "Who are the potential buyers or lease holders?- What crops would be grown on the land?- What would happen to the

---

[34] In relation to this it is relevant to mention what Bloch (1971) reports: *vazaha* (Europeans or foreigners) are seen with a mixed sense of scepticism, fear and envy, the same applies to the Malagasy who adopted western modes and wealth; a type of wealth, which is perceived as acquired by trickery rather than honest toil.

[35] During my stay in the country, I heard several comments about the anticorruption office BIANCO: how they could potentially decide whom to persecute or not persecute, upon their own judgment or advantage, and people should thus be careful.

[36] See Marcus (2004) for more examples.

[37] My translation from French to English.

crop? [...]- At what stage are the negotiations?- If an agreement has been reached or proposed, what are the proposed terms of the agreement?- What may happen to farmers or herders who live on or use the area?- Are there any proposed benefits to locals (such as new roads or employment opportunities)?- Are any groups advocating on behalf of locals in the negotiation process?" (Notes to Broadcaster 2008). These questions were on the minds of several Malagasy citizens and journalists after the Financial Times published its first article on the Daewoo deal. The answers from the government came late and were lacking transparency.

### *Economic Reasons*

"Malagasy already sold land to others in the past. Land here is a myth. A myth that allows to sell it more expensively", says a jurist from the MCI office for legal council (Carayol 2009). According to him, in Madagascar there is a political will to perpetuate the idea that ceding land to foreigners is not a normal market procedure, so prices can be raised in negotiations with international companies.

The economic factors play a central role in a land deal and economic discourses are given great attention. Profit is the primary goal of the investors, as well as a great concern for the hosting countries. The articles about the Daewoo land deal stressed the fact that apparently Madagascar would get very little out of the deal. This alerted Malagasy and fuelled the suspicions that only a few sitting in high positions were gaining personal advantages. A suspicion that given the past of the president's mixing of public and private affairs, was easy to sell to the public.

Nevertheless, economic reasons can go beyond the country's boundaries. Many articles supported the discourse that international interests were behind the large land deal: by denouncing Asian land grabbing some European countries, for instance, France, were protecting their own interests in Africa (Plasse 2009; Rakotomalala 2008). This was a discourse especially used by Ravalomanana and his followers (Plasse 2009).

### *Political Interests*

The deal with Daewoo became a rallying point for Rajoelina in late 2008 (Ploch 2009: 8). The media gave a great amount of coverage to the deal, while government denial never had the same attention. Rajoelina proficiently used this gap in his campaign against Ravalomanana (Rakotomalala 2008).

Rajoelina used the possible land transaction as a trigger to wake traditional feelings of attachment to land, through the large employment of populist discourses. In his speeches he preferred to use the word *tanindrazana,* which is a culturally meaningful concept; in the same way the term *selling* was a term preferred to the more exact leasing or rental; terms like neo- or new-colonialism were also largely used by Rajoelina and other exponents of his movement. During the analysis of the articles appearing in the press and my interviews it became clear that there was little discussion of the deal in terms of advantage/disadvantages, while a lot of coverage was given to the sin of ceding land to foreigners and to the lack of transparency showed by the government.

Nevertheless, after the shift of power and the official cancellation of the deal with Daewoo, Rajoelina's temporary government (High Authority of Transition), offered possibilities to other investments and welcomed farm ventures (Ramaro and Sguazzin 2009).[38]

## *Conclusion*

We have seen that discourses struggle for the production of meaning, which is a key to stabilising power relations (Jorgensen & Phillips 2002: 32). In other words, discourses are used to establish consensus and gain power. The analysis of the Daewoo deal shows how different stakeholders use discourses to establish meaning in their struggle to secure power. Through the analysis of the different discourses it is possible to understand the relevant points of a debate that was one of the main causes of Madagascar abrupt power shift.

As reported in the above paragraphs, discourses used by the political opposition as well as some newspapers presented land as ancestral. This is something many Malagasy are protective and sensitive about and as Dijk illustrates discourses are more persuasive if they use established beliefs (2001: 358). Some representatives of the Malagasy diaspora and bloggers largely used these discourses to express some of their concerns related to the deal.

---

[38] In Madagascar, other companies are trying to lease land, as for example the Indian Varun, as reported in Teyssier, Ramarojohn and Ratsialonana (2010).

Discourses developed by international organisations and NGOs showed concerns about the amount of land Daewoo tried to lease. This aspect was largely reported by international and Malagasy press, and was emphasised by the opposition to increase grievance toward the Ravalomanana government. International organisations and NGOs stressed in their discourses that leasing 1.5 million hectares of land to a foreign company could be seen as the beginning of neo-colonialism. The size was also too big to convince people that no evictions would be made to those lacking land titles. The extent of land at stake in the deal made the case very appealing for the media and exploitable by political forces in their discourses. Instead, lack of transparency of the deal helped fuel suspicions and disbelief, and served the political opposition campaign.

The real terms of the deal were never officially disclosed and the general impression was that the land would be leased in exchange for no payment.[39] All the negative aspects mentioned in relation to the size of the land as well as the lack of transparency were reinforced by the absence of compensation. NGOs, bloggers, as well as international and local press largely used this aspect in their discourses. In addition, the political opposition also stressed this point, awakening a sense of distrust toward the institutions. All these aspects were used to oppose the deal and the Ravalomanana government.

Above all, the bulk of the harvest was to be exported (Jung-a, Oliver and Burgis 2008). For the population of Madagascar, a country that is dependent on foreign aid to survive periods of food shortage, it is difficult to comprehend that food is to be exported.[40] The destination of the crops is one of the most striking aspects of this kind of deal (Notes to Broadcaster 2008), and this was largely used in discourses to oppose it.

Daewoo and Marc Ravalomanana's government underestimated a number of aspects when negotiating the lease of 1.3 million hectares of land in Madagascar. The economic discourses employed by supporters of the deal, together with the lack of transparency struggled against discourses featuring notions anchored in people's emotional concepts and traditional beliefs. The final outcome was a fierce protest that contributed

[39] Nevertheless, by employing 1.500 people, as prospected by Daewoo, the people obtaining a job would be one every 155 hectares, while the same amount of land could sustain a larger number than this (0.8 hectares/family), (Teyssier, Ramarojohn and Ratsialonana 2010: 7).

[40] WFP aids more than 500.000 people each year in Madagascar: Programme Alimentaire Mondial des Nations Unies – http://www.wfp.org/countries/madagascar.

to the country's power shift through what is internationally recognized as a *coup d'état*.

## *Bibliography*


Aliderson, F.A. (2004) Land as foundation of the Malagasy nation seen through Randria's "Kabary". Université d'Antananarivo. http://laboratoires.univ-reunion.fr/oracle/documents/375.html.

Andriamanana, V. (28/01/2009) Ivato: Dahio Corporation, rien a voir avec Daewoo Logistic, Midi Madagasikara. www.moov.mg (accessed 15/02/2009).

Andriamananoro A. (05.01.2009) Affaires Daewoo : Le « GASY-Pillage » rapide et durable de Madagascar sous le régime de Ravalomanana. www.Madagate.com (accessed 10/02/2009).

Astuti, R. & Harris, P.L. 2008. "Understanding Mortality and the Life of the Ancestors in Rural Madagascar". *Cognitive Science,* 713–740.

Bloch, M. 1971. Placing the dead tombs, ancestral villages, and kinship organization in Madagascar, London: Seminar Press.

Bloch, M. 1994. "The Slaves, the King, and Mary in the Slums of Antananarivo". In: *Shamanism, History, and the State.* Nicholas Thomas and Caroline Humphrey, eds. pp. 133–145. Ann Arbor: University of Michigan Press.

Bornschein, J. 2008. "La politique de Marc Ravalomanana à Madagascar". Dissertation / Travail de séminair, Grin.

Braun, J. v. & Meinzen-Dick, R. 2009. "'Land Grabbing' by Foreign Investors in Developing Countries: Risks and Opportunities". IFPRI.

Carayol, R. (24.02.2009) Madagascar: la braderie de l'île. Libération.

C'est du colonialisme et, à terme, une bombe à retardement. (09.12.2009) Ouest France.

Civil war looms' in Madagascar. (11.03.2009) BBC News.

Communiqué de presse de Daewoo Logistics. (10.04.2009) Orange Madagascar.

Corbett, C. (21.03.2009) "The DJ who has become Africa's youngest president". *Guardian.*

Cornwell, R. 2003. "Madagascar: Stumbling at the first hurdle?" *ISS Paper* 68, April 2003.

Cotula, L. (2007) "Legal empowerment for local resource control: Securing local resource rights within foreign investment projects in Africa." iied.

Cotula, L., Vermeulen, S., Leonard, R. and Keeley, J. (2009) Land Grab Or Development opportunity? Agricultural Investment And International Land Deals In Africa, IIED/FAO/IFAD, London/Rome.

CPJ (26.01.2009) Anti-government protests burns two TV stations. Committee to protect journalists. http://cpj.org/2009/01/two-tv-stations-burned-in-anti-government-protests.php (accessed 22/09/2009).

Daewoo fait main basse sur Madagascar (10.12.2008). tlaxcala.es. http://www.altermonde-sans-frontiere.com/spip.php?article8934 (accessed April 30, 2009).

Daewoo Logistic Corporation (21.11.2008) Press Release. http://www.maep.gov.mg/dossiers.html (accessed June 08, 2009).

Dahl, Ø. 2008. *Merkeverdige Madagaskar. Øya mellom øst og vest.* Oslo, Spartacus.

De Schutter, O. (11.06.2009) "Large-scale land acquisitions and leases: A set of core principles and measures to address the human rights challenge", UNHRC.

Délivré, A. (1974) L'histoire des rois d'Imerina: interprétation d'une tradition orale, Nanterre, Société d'ethnologie.

Démenti de Daewoo: aucun acte d'achat signé, aucune expropriation en vue. (02.02.2009) www.topmada.com (accessed 06/06/2009).

Dijk, T.A. van, 2001. "Critical Discourse Analysis". In D. Tannen, D. Schiffrin, and H. Hamilton (eds), *Handbook of Discourse Analysis.* Oxford: Blackwell, 352–371.

Ellis, S. 1985. *The rising of the Red Shawls: a revolt in Madagascar, 1895–1899*, Cambridge, Cambridge University Press.

EU to keep Madagascar aid block (4 June 2010) BBC News.

Evers, S.J.T.M. 2006. "Expropriated from the Hereafter: The Fate of the Landless in the Southern Highlands of Madagascar", *Journal of Peasant Studies*, vol. 33, nr. 3, pp. 413–444.

Evers S.J.T.M., van den Haak M., Lingnau I., Lokhorst N. and Pronk C. (01.09.2006) National Legislation and Local Practices: Competing Jurisdictions in Land Management in Madagascar. *Taloha*, n.16–17 http://www.taloha.info/document.php?id=336.

Evers, S.J.T.M. (2002) "Constructing history, culture, an inequality: the Betsileo in the extreme Southern Highlands of Madagascar", Leiden: Brill Academic Publishers.

FAO. 2009. Economic and Social Development Department (June 2009) From Land Grab to Win-Win, Policy Brief.

Felana (14.01.2009) Le Mrfdat révèle la vérité sur l' "affaire" Daewoo Logistics – Que cessent les mensonges er rumeurs. Le Quotidien. Foncier. La presse réunionnaise évoque l'affaire Daewoo. (30.12.2008) www.moov.mg (accessed February12, 2009).

Fournet-Guérin, C. « La dimension spatiale du fait national à Madagascar : un État-nation en Afrique ? », L'Espace Politique 7 | 2009-1. http://espacepolitique.revues.org/index 1244.html.

Galibert, D. (2009) Les gens du pouvoir à Madagascar - Etat postcolonial, légitimités et territoire (1956–2002), Antananarivo, Karthala Editions.

Grabbing land for food. 2009. GRAIN. www.grain.org (accessed 01/10 2009).

Graeber, D. 2007. Lost people magic and the legacy of slavery in Madagascar, Bloomington, Ind., Indiana University.

Hanson, P.W. 2007. "Governmentality, Language Ideology, and the Production of Needs in Malagasy Conservation and Development." *Cultural Anthropology* 22(2): 244–284.

Heuvel M. et S.J.T.M. Evers. (20.11.2007) Once There Were Rice-fields… Life Away from the Ancestral Land. *Taloha*, n.18, http://www.taloha.info/document.php?id=575.

Hogg, J. 2009. Madagascar's "militant mayor". BBC News.

Hundreds protest Madagascar Mayor's sacking. (04.02.2009) Agence France-Presse.

Investissement agricole. Les terres pour les Sud-Coréens continuent à faire des vagues. (25.11.2008) www.moov.mg (accessed 12/02/2009).

Jaovelo-Dzao, R. (1996) Mythes, rites et transes à Madagascar. Paris, Karthala.

Jørgensen, M. & Phillips, L. (2002) *Discourse analysis as theory and method.* London Thousand Oaks, Calif., Sage Publications.

Journée foncier. 2009. CIRAD meeting. www.cirad.fr (accessed June 10, 2010).

Jung-a, S., Oliver, C. and Burgis T. 2008. Daewoo to cultivate Madagascar land for free. *Financial Times.*

Keller, E. 2008. "The banana plant and the moon: Conservation and the Malagasy ethos of life in Masoala, Madagascar." *American Ethnologist*, 35(4): 650–664.

Kraemer, A. 2007. "Voices, Counter-Voices, Silences: Contested Representations of a Mining Project in Madagascar". Unpublished MA Dissertation, School of Oriental and African Studies, London.

L'opposition porte plainte. (18.12.2008) *Midi Madagasikara.*

Land grabbing in Africa: an overview. (08.06.2009) African farms news in review, n. 69. Farm Radio weekly: http://weekly.farmradio.org/?p=1004 (accessed 09/09/2009).

Li, T.M. 1996. "Images of Community: Discourse and Strategy in Property Relations", *Development and Change* 27: 501–527.

M.R. and E.R. (2009) Le futur… entre leurs mains. Revue de l'Océan Indien. N. 298.

Mackenzie, D. 2008. Rich countries carry out '21st century land grab', *New Scientists.*

Madagascar Action Plan (MAP) 2007–2012, commitment 4, challenge 1.

Madagascar leader axes land deal (19.03.2009). BBC News.

Madagascar President forced out (17.03.2009). BBC News.

Madagascar sentences ex-president (03.06.2009). BBC News.

Madagascar: l'affaire Daewoo alimente la colère contre le gouvernement. (10.02.2009) Agence France-Presse.
Madagascar: Speech of His Excellency Marc Ravalomanana. (31.03.2009) allAfrica.com (accessed 09/09/2009).
Maka, A. 2008. "Madagascar, l'affaire Daewoo confirmée". *La Vérité.*
Makunike, C. 2009. "The lessons of Daewoo's failed Madagascar land lease deal, African Agriculture". On http://farmlandgrab.org/ (accessed 08/06/2009).
Malagasy land and foreign investments. (21.10.2008) Blog. http://andrydago.wordpress.com/2008/10/21/malagasy-land-and-foreign-investments/#comments (accessed 10/02/2009).
Marc Ravalomanana, un voyou sans foi ni loi (20.11.2009). *Soa Madagasikara.* www.madagasikara-soa.com (accessed 05/10/2009).
Marcus, R.J. 2004. Political Change in Madagascar: Populist democracy or neopatrimonialism by other name? ISS.
McGreal, C. 2009. "African Union suspends Madagascar over 'coup'". *The Guardian.*
Müller D.F. & Evers S.J.T.M. 2007. "Case Studies of Land Access Practices in Fianarantsoa Province, Madagascar". *Taloha* n.18, http://www.taloha.info/document.php?id=589.
Navalona, R. 2009. "Projet Daewoo: pas d'expropriation de terrain en vue!" *Midi Madagascar* n.7738.
Notes to broadcasters on foreign-owned farmland (22.12.2008), Farm Radio Weekly. www.farmradio.org (accessed 11/02/2009).
Opinion divided over Ravalomanana. 2006. IRIN.
Pellerin, M. 2009. "Madagascar: un conflit d'entrepreneurs? Politique Africaine", p.152–165.
Plasse S. 2009. "Madagascar: Why is Marc Ravalomanana accusing France? The Ex Madagascan president criticizes French support for his rival". Afrik-news.
Ploch, L. 2009. "Madagascar's 2009 Political Crisis". Congressional Research Service.
Programme Alimentaire Mondial des Nations Unies – http://www.wfp.org/countries/madagascar.
Pronk C. & S.J.T.M. Evers (20.11.2007) "The Complexities of Land Access in Southeast Madagascar". TALOHA n.18, http://www.taloha.info/document.php?id=562.
Rajoelina demande la destitution de Ravalomanana et se proclame responsable de l'exécutif. (31.01.2009) *Madagascar Tribune.*
Rakotoarivony C.B. & Ratrimoarivony, M. 2006. "Terre ancestrale ou parc national? Entre légitimité sociale et légalité a Madagascar". *Vertigo*, 17.
Rakotomalala, L. 2009. Madagascar: South Korean Land Deal Spark Controversy, Global Voices Online.
Ramambazafy, J. 2009. "Madagascar: pourquoi la transition? La constitution Ravalomanana et la notion de "Tanindrazana"". www.Madagate.com (accessed 10/02/2009).
Ramaro O. and Sguazzin A. 2009. Madagascar may welcome farm ventures after Daewoo cancellation. Bloomberg.
Ramiakajato, A. & Ravelomanantsoa, S. (2009) L'Affaire Daewoo: mythes et égalités. *Revue Juridique et Fiscale de MCI*, n.46.
Razafiarijaona, J. 2009. "Bilan diagnostic de sécurisation foncière et perspective pour Madagascar." *Revue Juridique et Fiscale de MCI*, n. 45.
Razafiarijaona, J. 2009. "Bilan diagnostic de sécurisation foncière et perspective pour Madagascar." *Revue Juridique et Fiscale de MCI*, n. 46.
Ryall J. and Pflanz M. 2009. "Land rental deal collapses after backlash against 'colonialism'", *Telegraph.*
SADC does not, cannot recognize Rajoelina. (19.03.2009) Independent Online. *Agence France-Presse.*
Shepard, D. & Anuradha, M. (2009) The Great Land Grab. Rush for World's farmland Threatens food security for The Poor, The Oklahoma Institute.
Smaller C., & Mann., H. 2009. "A Thirst for Distant Land. Foreign investment in agricultural land and water". International Institute for Sustainable Development. IISD.

Smith, R., 2003. "Land Tenure Reform in Africa: a Shift to the Defensive". *Progress in Development Studies*, 3(3): 210–222.
Terres Malgaches. www.terresmalgaches.info (accessed 10/10/2009).
Teyssier A., Raharison, H. & Ravelomanatsoa, Z. 2007. "La reforme foncière de Madagascar ou le pari de la compétence locale." in *FAO (Ed.) Reforme Agraire, Colonisation et Coopératives Agricoles*.
Teyssier A., Ramarojohn L. and Andrianirina Ratsialonana R., "Des terres pour l'agro-industrie internationale ? Un dilemme pour la politique foncière malgache", *EchoGéo*, Numéro 11, 2010.
Troops oust Madagascar minister. (10.03.2009) BBC News.
Tuquoi J.P., Messmer P. and Hervieu S. (2008) La Corée du Sud relance la course aux terres agricoles http://www.lemonde.fr/planete/article/2008/11/20/la-coree-du-sud-relance-la-course-aux-terres-agricoles_1120902_3244.html (accessed 08/06/2009).
US calls Madagascar change a 'coup', suspends aid. (20.09.2009) *Agence France-Presse* (Google News). (accessed 01/10/2009).
Veum, B., 2010. "The role of the Constitutional Court of Madagascar in the 2009 political crisis". SIK/Université de Toliara (Unpublished paper).
Vinciguerra, V. 2009. "Madagascar: waiting and uncertainty follows 'Red Saturday'". www.afrika.no. (accessed 10/04/2009).

# INDEX

www.ingramcontent.com/pod-product-compliance
Lightning Source LLC
LaVergne TN
LVHW011122300926
849160LV00008B/480
* 9 7 8 9 0 0 4 2 5 6 2 2 4 *